ERROR AND UNCERTAINTY IN DIAGNOSTIC RADIOLOGY

ERROR AND UNCERTAINTY IN DIAGNOSTIC RADIOLOGY

MICHAEL A. BRUNO, M.D., M.S., F.A.C.R.
Professor of Radiology & Medicine
Vice Chair for Quality & Safety
Chief of Emergency Radiology
Penn State Health/Milton S. Hershey Medical Center
and *The Penn State College of Medicine*
Hershey, Pennsylvania

OXFORD
UNIVERSITY PRESS

OXFORD
UNIVERSITY PRESS

Oxford University Press is a department of the University of Oxford. It furthers the University's objective of excellence in research, scholarship, and education by publishing worldwide. Oxford is a registered trade mark of Oxford University Press in the UK and certain other countries.

Published in the United States of America by Oxford University Press
198 Madison Avenue, New York, NY 10016, United States of America.

Library of Congress Cataloging-in-Publication Data
Names: Bruno, Michael A., author.
Title: Error and uncertainty in diagnostic radiology / Michael A. Bruno.
Description: New York : Oxford University Press, [2019] | Includes index.
Identifiers: LCCN 2018039925 | ISBN 9780190665395 (POD)
Subjects: | MESH: Radiography | Diagnostic Errors |
Diagnostic Imaging | Case Reports
Classification: LCC RC78.7.D53 | NLM WN 200 | DDC 616.07/ 572—dc23
LC record available at https://lccn.loc.gov/2018039925

Cover Concept: Andrea L. Sakse, M.D.

This book is dedicated to my wife (of more than 30 years!) Andrea, who provided me with needed insights, encouragement and ongoing support during this long project, and most importantly provided me with the benefit of her amazing time-management skills in the final months of writing.

In doing so, she has made this book possible.

CONTENTS

SECTION 2 THE SCIENCE OF IMAGING AND PERCEPTION AND THE PROBLEM OF ERROR IN RADIOLOGY

SECTION 3 STRATEGIES FOR REDUCING RADIOLOGICAL ERROR

FOREWORD

BY MARK L. GRABER, M.D.

"The most important doctor you've never met . . ."

Diagnostic errors have always been with us, but mostly in the background. All physicians know about the problem from the ever-present threat of a malpractice suit, but in their everyday lives, it seems pretty remote. If you ask their honest opinion, most physicians would probably say that almost all of their diagnoses are correct.

We now know that probably isn't true. According to our best estimates from a wide range of research approaches, probably 10% or more of diagnoses in everyday practice are wrong, are missed entirely, or should have been made much earlier. Dr. Hardeep Singh, a clinical investigator in the Department of Veterans Affairs and Baylor Medical Center, found that one in 20 primary care patients will experience a diagnostic error every year. From autopsy studies, even with all our modern technology, major discrepancies are found in 10% to 30% of cases, adding up to some 40,000 to 80,000 deaths a year. The widely respected ECRI Institute designated diagnostic error as the number-one safety concern in 2018, beating out opioid addiction, health IT problems, and other classic concerns. The National Academy of Medicine tells us that "each of us is likely to experience one or more diagnostic errors in our lifetime, sometimes with devastating consequences."

That's the bad news. The good news is that the problem finally has emerged as one demanding our attention. And that brings us to this book, *Error and Uncertainty in Diagnostic Radiology*. Dr. Michael Bruno's book is all about radiology, and more specifically about the quality and safety of medical imaging. If we want to minimize harm from diagnostic error, we first have to understand the diagnostic process itself, how quality and safety evolves in this process, and how diagnostic errors arise and can be prevented. Through the lens of radiology, Dr. Bruno explores each of these areas, providing us with a candid, refreshing, and insightful look at the diagnostic process in this field from start to finish. When I take my car in for an oil change, I can see exactly how the job gets done. I can see how the mechanic drains out the ugly old oil and replaces it with new oil, golden and so smoothly flowing, one can at a time. Understanding how the radiologist's job gets done is quite another matter; it's cognitive stuff. The beauty of Michael Bruno's writing is his ability to open up this invisible world for us so that we can see what happens behind the curtain, or between the ears in this case, and begin to understand how a radiologist arrives at a diagnosis and where and how errors creep in. Anyone interested in diagnostic error and reducing it would benefit from this firsthand review, presenting Dr. Bruno's experience and insights as a clinical radiologist and academic investigator and, by the way, a very entertaining writer.

Compared to the 10% to 15% error rate in frontline specialties like internal medicine, pediatrics, and emergency medical care, radiology fares much better, with an estimated error rate thought to be in the range of 2% to 4%. But with the millions of imaging exams conducted every day in our country, that's not exactly comforting news. A highlight of this book is Dr. Bruno's elucidating for us a key aspect of the radiologists' skillset, namely their ability to find the abnormalities on an imaging study and characterize them. The growing understanding of perception is fascinating and opens the door to interventions to prevent these errors, where a key finding on an image is simply missed.

Of course, the low-hanging fruit for reducing harm from radiology-related errors lies in the system-related breakdowns. Many of these occur at the interface between radiologists and their "customers," the clinicians ordering and acting on requested imaging examinations. The legion of errors in this domain is large. Sometimes the clinician doesn't know the best test to order for a given complaint and chooses something less suitable; or a report just gets lost and never reaches the clinician; or an incidental finding on the imaging exam is detected but no one follows up on it; or suggestions for additional imaging studies are ignored. Many of these issues can be addressed by the radiology practice acting on its own, but other solutions will require much better communication between the radiologist and the clinician. Making the patient a part of the diagnostic team also has great potential to improve the diagnostic process that involves imaging. For example, giving patients the results of their imaging examinations could virtually eliminate those breakdowns where test results just "fall between the cracks." The evidence for this is convincing. This problem has virtually vanished when it comes to mammography, thanks to the federal Mammography Quality Standards Act (MQSA) requirements that patients receive a written summary of their test results, in lay language.

Although the book stays focused on the diagnostic process in radiology, understanding error in medical imaging is an excellent paradigm for understanding diagnostic error in general. The system issues, the cognitive shortcomings, and the inescapable uncertainty that exists at every step of diagnosis in radiology echo the exact same concerns in every other discipline of medicine.

A valuable contribution of this book is its classification of these issues in a novel and easy-to-remember framework based on the three major elements that underlie an understanding of the diagnostic process: *how we work, how we think*, and *how we are made*. Another very likeable feature is how Dr. Bruno uses both "Safety 1" and "Safety 2" perspectives. Safety 1 is the concept that learning from errors is the key to improvement. Safety 2 refers to the learning that emerges from studying when things go right, as they most typically do in radiology. Yes, we would love to reduce the 2% to 4% error rate to some lower range. But it is indeed remarkable, given the complexity of image interpretation and the essentially infinite amount of information contained in a single imaging study, how often the correct diagnosis emerges in the radiologist's mind.

Some definitions of a profession include the notion that its members are proactive and highly concerned about improving the quality of their work. *Error and Uncertainty in Diagnostic Radiology* is an important contribution to improving the quality and safety of medical imaging and represents an elegant example of how a single individual, in this case one of the profession's leading researchers and educators, can impact an entire field. To its great credit, the larger radiology community is also actively engaged in quality improvement efforts. The American College of Radiology leads this work, and their many projects directed at improving diagnosis are listed on their website. This emphasis on improving diagnosis is exemplary; except perhaps for the work done by clinical pathologists, no other profession has devoted the same time and effort to this cause.

Radiologists are special people. They are happy to sit in a dark room day after day, by themselves, striving to help patients they have never met and will never touch. But they care very deeply about patients and know that the quality of their work is critically important to patients' well-being. Dr. Bruno cites Drs. Bruce Hillman and Jeff Goldsmith for describing radiologists as "the most important doctor you've never met." I completely agree, and we can thank them for the quality of the service they provide to each of us as patients. And a special thanks to Michael Bruno for this book, which has the potential to stimulate a great deal of thought and hopefully action to improve the safety of medical imaging, and diagnostic safety more generally, in the process.

Mark L. Graber, M.D., F.A.C.P.
Professor Emeritus of Medicine, SUNY at Stony Brook
Founder and President,
Society to Improve Diagnosis in Medicine

FOREWORD

BY STUART J. FIRESTEIN, PH.D.

I have two favorite quotes regarding the importance of uncertainty and not knowing. One is by the poet John Keats and appeared in a letter to his brother. In it he coins the term "negative capability," which he defines as "when a man is capable of being in uncertainties, mysteries, doubts, without any irritable reaching after fact and reason." The second is from the physicist Erwin Schrodinger, who noted that "In an honest search for knowledge you quite often have to abide by ignorance for an indefinite period." Both of these quotes suggest that ignorance and uncertainty, doubt and failure, are not to be avoided so much as accepted. Both recognize that these states, possibly uncomfortable (thus Keats suggesting that one must avoid irritability), are nonetheless a creative state, a time when new ideas arise, when the light suddenly breaks through on the previously unsolvable. The interesting thing about these two quotes is that one is from an artist and the other from a scientist. Medicine, it is often noted, is a blend of art and science.

The notion of negative capability that Keats introduces is one worth inspecting. At first it may seem contradictory. Capability and negation don't usually partner. But developing a patience for the unknown is special too; the ideal state for the creative mind, as Keats called it. Patience with ignorance, rather than frustration, the capability to see the uncertain as the possible, the mystery as opportunity—these are fertile places for the inquiring mind.

As for error, the second great topic of this necessary and overdue book, I will quote Benjamin Franklin, arguably America's first scientist. "Perhaps the history of errors of mankind, all things considered, is more valuable and interesting than that of their discoveries. Truth is uniform and narrow . . . but error is endlessly diversified."

All of this may be good philosophy, but aren't there often disastrous consequences of uncertainty, and especially error, in medicine? It's well and good in research and art to court failure and doubt, to use them to power your creativity and motivate your problem solving. But with an ill or injured person facing you, the potential harm that can result from an error must give us pause. It is important here to distinguish between an error born of genuine uncertainty and lack of data rather than mere sloppiness and ineptitude. The second is preventable and inexcusable. But the errors that are endemic to the practice of the inexact science and art of medicine are not a cause for recrimination, or litigation; they are a piece of what we should recognize as a remarkably successful therapeutic practice. Uncertainty and error cannot be simply amputated from medicine as if it were an offending limb or a diseased organ.

How, then, should it be practiced or become part of the practice of medicine? There is no simple answer to this question, or there would be no book following this foreword. There are a few prescriptive actions that can be taken, but overall it seems to require a reconsidering of the very practice of medicine and, perhaps as importantly, the way physicians are educated. The Harvard Medical faculty has a regular series of seminars and symposia on reforming medical education. It moves somewhat glacially, but the issues are at least addressed and brought to the forefront.

One of the most pervasive of these issues is what the patient expects from the doctor. This relationship is in many ways the key obstacle to doctors allowing error and uncertainty to become a part of their practice. Most doctors feel a social pressure from their patients to be prescriptive, to tell them what's wrong and what they can do about it. And even if there is nothing to do about it, at least tell me what's wrong. And so doctors often pretend to be more sure than they are, more certain about what to do. The difficult thing here is that this often works. The well-known placebo effect, where simply being under the care of a confident doctor will make you well, regardless of what she does, including nothing, suggests that we should keep this ruse up.

And this is the continual tug, the tension in medicine, a practical art and sophisticated science, between patient care and understanding disease. The doctor of today may well be called upon to act confident and certain with her patients, while developing means of examining error in a guilt-free, blameless environment, a place where error and misjudgment can be shared and tested and experimented upon while we go ahead and treat patients as best as we can. The good news is that for all the errors, "as best as we can" is pretty good—especially compared to the shaman who was dead sure of his concocted cure.

As a document of error and uncertainty this book fills a niche that could allow those medical faculties that are serious about reforming education to embrace the failures and doubts, and the working members of the profession to recognize the importance of error in their daily patient contacts and feel no shame for it. This book takes you deep inside the art and practice of medicine, radiology in particular, for a penetrating case history that we can use to come to grips with the realities of medical practice, and begin thinking of ways to improve it—while doing no harm to it.

Stuart J. Firestein, Ph.D.
Professor, Department of Biological Sciences
Columbia University
New York, New York

INTRODUCTION

Diagnostic radiology is a specialty of medicine that is expressly dedicated to the diagnostic process. Few patients ever meet the radiologists who contribute to their care, as radiologists practice mostly "behind the scenes," serving primarily as consultants to other physicians. As a result, most patients are unaware of how radiologists are involved in their healthcare and may be surprised to see radiologists' interpretive fees on their medical bills. Radiologists' connection to the patients they serve is not as close or immediate as it is for other physicians. Few patients feel any particular attachment to their radiologist, nor would they mention "my radiologist" in a conversation about their recent trip to the hospital, although they might mention "my cardiologist" or "my surgeon." Yet the role of radiologists is increasingly gaining in impact, the result of progressive trends toward greater utilization of radiology tests along with a diminishing role for the physician history and physical examination in the diagnostic process. The role of radiology continues to expand, and in recent years it has become paramount in the evaluation and diagnostic formulation for almost every hospitalized patient and almost everyone seeking care in hospital emergency rooms throughout the developed world.

Pressures on doctors and healthcare organizations for increased efficiency have led to changes in the process by which doctors evaluate patients' problems, with a clearly decreasing emphasis on the time-consuming and laborious history-taking and comprehensive physical examination and a much greater, and growing, reliance on advanced medical imaging, especially computed tomography (CT) and magnetic resonance imaging scans. Some of this change is being driven by concerns for speed—the added cost of a CT scan is offset by speeding a patient's "throughput" in the emergency department, for example—and some of the change is driven by a desire on the part of physicians for a higher level of diagnostic certainty before making any therapeutic decisions. The practice of "defensive medicine," whereby physicians, especially in the emergency setting, seek greater objective evidence from testing to support their actions in hope of reducing the risk of a lawsuit, is also a contributing factor, although recent studies have suggested that defensive medicine is not as large a driver of imaging overutilization as is commonly believed.

Each modern high-tech imaging examination produces hundreds of detailed images, which are constructed and processed in myriad ways, resulting in a very complex dataset for which essentially all clinical physicians require the help of a trained imaging specialist to interpret. Those radiological interpretations are themselves being done under greater time pressure than ever before, because, as I've alluded to, imaging is increasingly becoming the principal means of patient assessment in the acute setting, and the volume of studies requiring interpretation has grown much faster than the radiologist workforce. The result is that each radiologist (on average) must interpret more studies during each hour of each workday than ever before. In the emergency department setting, for example, the utilization of advanced imaging, particularly CT scanning, has increased more than 300% in recent years, and in most centers, including the one in which I practice, the expectation is that imaging studies will be interpreted very promptly by a radiologist, with a written report available to the emergency physician within an hour, if not sooner.

Drs. Bruce Hillman and Jeff Goldsmith, in their excellent 2011 book *The Sorcerer's Apprentice*, referred to the radiologist as "the most important doctor you've never met." They noted that most patients don't even realize that their imaging studies are interpreted primarily by radiologists rather than by their own physicians or surgeons. In fact, multiple published studies have shown that around half of patients surveyed (in both the United States and in Europe) were not even aware that radiologists are physicians, let alone that the specialists Hillman and Goldsmith refer to as "these shadowy figures" must spend 5 to 7 years after medical school in focused, intensive study as well as long days and many nights in supervised practice (residency), which is essentially an apprenticeship under the supervision of experienced teaching radiologists, with gradually increasing degrees of autonomy and independence. This training culminates in a multiyear process in which radiology residents must ultimately demonstrate their knowledge and proficiency in a series of examinations administered by the American Board of Radiology before they are certified to practice independently.

During this training, a radiology resident will likely have interpreted on the order of 60,000 imaging studies under varying degrees of direct supervision, and the majority develop considerable expertise.

Dr. Robert Wachter, in his insightful book *The Digital Doctor*, published in 2015, has written eloquently about the demise in recent years of "radiology rounds," a ritualistic practice in which teams of physicians would consult in person with radiologists on a daily basis within the radiology department, reviewing studies performed over the prior 24 hours. I can speak from my own experience that such daily consultation with teams of physicians was a very rewarding and satisfying experience for all concerned; it was still the mainstay of radiological practice when I began my clinical career in the early 1990s. The give-and-take conversations I would have throughout the day with various teams of clinicians (including medical students, residents, and their supervising attending physicians) allowed us all to think through the nuances of how each patient's imaging findings meshed with his or her symptoms, family history, and physical exam findings, and I am convinced it improved the accuracy of my radiological diagnoses. It is but is a distant memory now, as busy clinicians prefer to read radiologists' final, dictated reports "on the fly," often while sitting with the patients in clinic, and typically within minutes of the radiologist electronically signing each report throughout the day. This fundamental change in the style and pace of radiology practice, including how doctors utilize radiologists' services, how radiologists' written reports are structured and used, and especially how the discipline is taught to trainees, has far-reaching ramifications for patients and for doctors.

Since the practice of radiology has become such a large contributing factor to the overall diagnostic process for almost every patient in the developed world, it stands to reason that it should also make a large contribution to the problem of diagnostic error, a topic of increasing worldwide concern. In 2015, the National Academy of Medicine published a landmark report, *Improving Diagnosis in Health Care*, highlighting the importance of diagnostic error as a cause of patient harm and estimating that the prevalence of such error is high enough that every American is likely to experience at least one diagnostic error in his or her lifetime. While the extent that errors in radiology underlie the overall problem of diagnostic error cannot be known precisely, it is likely to constitute a substantial contribution to the overall problem of diagnostic error. It thus behooves us to develop a better understanding of what constitutes an "error" in radiology, as well as to develop a taxonomy of error as applied to radiology, and also try to understand the mechanistic underpinnings of radiologist error to the extent these are currently knowable.

It can be said that error in radiology may represent a microcosm of the more general problem of diagnostic error overall, including systems-based complexities, communication lapses between caregivers, cognitive lapses due to the effects of biases and faulty reasoning, and perceptual failings. As a specialty, radiology and radiologists have been willing, at least since the contributions of Dr. Leo H. Garland in the 1940s and 1950s, to turn the focus on ourselves and ask ourselves tough questions about our own efficacy, accuracy, and diagnostic performance, including when we err, in systematic ways—with the overall goal of improving our own diagnostic accuracy.

But first we must understand, at the most fundamental level, the essence of what radiology practice is and what it contributes to medicine. I believe that radiology is primarily a means to reduce (but rarely to eliminate) diagnostic uncertainty. In my view, the entire process of radiological imaging, up to and including radiologists' interpretation, accomplishes nothing more than to reduce the degree of uncertainty present such that a treating physician will have sufficient confidence in taking, or withholding, action. This is essentially the definition of a "measurement" in the context of Claude Shannon's information theory, and it presupposes that a high enough degree of uncertainty exists prior to turning to radiology that a treating physician would be uncomfortable making the decision to either undertake or forgo any given available treatment.

In his seminal 2016 treatise on the role of error and failure in advancing science, entitled *Failure: Why Science Is So Successful*, Professor Stuart Firestein of Columbia University, an eminent biologist, professor, and essayist on the scientific process, makes the point that the experience of failure itself, especially when followed by a process of self-reflection and careful analysis, ultimately culminating in each scientist learning from his or her own (and other's) failures, has led inexorably to an increase in the depth of scientific knowledge and improved scientific performance. He argues that episodic failure is therefore fundamental to all progress in the sciences. Making an analogy to medicine in a section of his text entitled "Failure in the Clinic," he ends by saying (p. 143):

> I don't want to dwell too long on medicine because it is so fraught with moral and ethical issues. Not to mention the

> economics that has come to drive so much of it. I refer not only to physicians, but even more to the entire medical infrastructure—from hospitals to insurance companies to device and instrument makers to the drug business. Medicine is a science, an art, and I'm afraid now an industry. It clearly deserves its own book on failure, but I am not the person to write it.

This book attempts to rise to Dr. Firestein's challenge, at least for the specialty of diagnostic radiology. In it, I hope to confront the intensely uncomfortable topic of clinical error in radiology in a rigorous, unblinking, but ultimately optimistic way. This includes exploring the mechanistic (reductionist) analysis of the causes (or potential causes) of radiologist errors as well as attempting to provide a more holistic understanding of these errors in the broader context of human error and fallibility and the high level of uncertainty inherent in the process itself. My aim is to clarify, and at least partially disentangle, the many complex variables that contribute to the overall problem of radiological error. We will analyze the various types and classifications of radiologist error, their relative prevalence, and their possible causes. In the process, we will also explore some related issues, such as the many relevant scientific experimental results in human visual and perceptual physiology, and current thinking regarding radiologists' accountability for error at multiple levels, including professional peer review, professional self-regulation, and medical malpractice litigation. We will also explore the multiplicity of hopeful error-reduction strategies being evaluated in current research and consider the history of unsuccessful error-reduction strategies that have been employed in recent decades to reduce radiologist error but have thus far failed to effect any meaningful performance improvement. Understanding our past failures should help to drive forward progress, as it has done in other fields of science and medicine.

The final chapter is forward-looking and covers the broad topic of artificial intelligence (AI), machine learning, and deep learning as applied to the goal of providing meaningful error reduction and mitigation of harm for patients, including a review of current research in this area. This chapter also gives some consideration of the broader future impacts of AI on the field of diagnostic radiology.

This book is thus organized into a handful of overarching themes: (1) an assessment of the problem of error in the practice of diagnostic radiology, including placing the problem into the broader perspective of the underlying uncertainty that is present, (2) providing a taxonomy and epidemiology of error in radiology, with analysis of the many contributing and confounding issues that contribute to the problem of error in radiology, (3) a discussion of accountability for error and strategies for dealing with radiological error after the fact, (4) a review of related science in human visual perception and cognition and their potential contribution to a broader "theory of error" in radiology, (5) a look at research and practical strategies for error reduction and early detection, and (6) a forward-looking roadmap for the future of the field that includes AI, and what a future "post-AI" world of radiology might mean in terms of error reduction.

At the beginning of the major chapters you will find carefully selected "clinical vignettes of error," which are provided with the goal of placing the issue of radiology error and uncertainty more firmly into its human, clinical context. Our errors have consequences, which is what makes our efforts to understand and reduce error in radiology so vitally important. These vignettes are fictional but are very similar to real-life events I've encountered in my own practice over the past 30 years.

I am deeply indebted to many colleagues in radiology, medicine, and the sciences basic to medicine, including those who have truly been mentors to me, including the late Professor Eric N.C. Milne, whose early mentorship (now more than 34 years ago) convinced me to pursue a career in radiology, as well as to several colleagues whom I consider to be true friends, and those who are both true friends *and* true mentors, such as Drs. Timothy J. Mosher, Howard Egeth, Elizabeth A. Krupinski, Martin G. Pomper, Charles E. Kahn, Frank J. Lexa, Leonard Berlin, and Hani H. Abujudeh, among others. To each, I am deeply indebted for their friendship, strong support of this project, as well as their cumulative equivalent of many lifetimes of careful research that I have attempted to summarize in this slender volume—as well as for their providing critical readings, thoughtful discussions, and many valuable suggestions throughout the course of the writing. I am also especially grateful in that regard to Dr. Mark Graber, founding president of the Society to Improve Diagnosis in Medicine (SIDM), both for his comprehensive editorial feedback and guidance throughout this project and also for providing his rich perspective on the topic of diagnostic error, as well as to my many friends and colleagues in the SIDM, especially including the SIDM's Board for their support of this project from its earliest inception.

I also wish to thank my executive publisher at Oxford Press, Andrea Knobloch, who believed in this project from the outset and has tirelessly helped me to bring it to fruition.

This is the second book project in which I've had the privilege of collaborating with Andrea, which is a testament to her patience and forbearance.

Finally, I wish to thank and acknowledge my close friend and Penn State Hershey colleague Dr. L. Scott Winner, an award-winning teacher of medical humanities and himself an accomplished writer of medical creative nonfiction, for many fruitful hours of discussion over the past decade on how we (individually as human beings, and as a profession) need to face up to, and deal with, our own mistakes.

SPECIAL ACKNOWLEDGMENT

This book project spanned more than a year, and in it I've attempted to summarize a growing understanding of the problem of error in radiology that has developed slowly over several decades. In writing this book I have benefited from the insight, help, and guidance from a large number of people, including those whose work I have cited in the text. The most influential of these have already been named.

I would like to also formally acknowledge the involuntary contribution of a very specific group of people who are the real reason this book was written: the many patients who have been harmed by the mistakes of radiologists, including (and especially) my own. While the harm caused by diagnostic errors cannot always be undone, I hope that there is some comfort in knowing that we radiologists are truly committed to learning from our mistakes, understanding more deeply why they happened, and, most importantly, improving our performance so that we can do better by you the next time our paths cross.

SECTION 1

PARSING THE SHADES OF GRAY

Error, Uncertainty, and the Goals of Radiology

1

UNCERTAINTY AND INTERPRETIVE ERROR

A 34-year-old man comes to the emergency department (ED) complaining of severe right flank pain. He is questioned, and examined. He is found to be apprehensive, uncomfortable, a little sweaty, and restless. He says the onset of symptoms was sudden. He has no fever. The left flank, over the kidney region, is tender to palpation. The ED physician suspects a kidney stone and obtains a non-contrast computed tomography (CT) scan of the abdomen and pelvis. The CT scan confirms the presence of a small, partially obstructing stone within the mid right ureter, with mild backup of urine in the right kidney, which is swollen as a result. The size of the stone is such that it ought to be able to pass on its own, although the process is likely to be painful for days to come. The patient is treated for pain and advised to keep well hydrated and to strain his urine to collect the stone when it passes. But the radiologist failed to notice some small, nondescript (and nonspecific) mesenteric nodules clustered in the left upper abdomen, which were far from the area of interest. The patient is sent home and does well. Two years later he is again seen in the ED, with new symptoms of unexplained weight loss, bowel irregularity, and abdominal pain that have been gradual in onset. A new CT scan shows large, malignant-appearing masses where the small, overlooked nodules had been before, and now there are metastatic lesions in the liver and lungs. A subsequent biopsy confirms the diagnosis of metastatic cancer. It seems likely in retrospect that the cancer was present but was missed on his CT scan from two years earlier; had the finding been diagnosed then, before the tumor had grown and spread, there would have almost certainly been a better outcome, including a longer survival, although the kind of cancer present is generally not curable, even if caught early.

Radiology is a medical specialty that is practiced in an environment of extraordinarily high uncertainty. Many patients are unaware of what radiologists do, are not aware of the contributions radiologists have made in their own medical care, and most likely have never met "their" radiologist. Historically, the radiologist has been the "doctor's doctor," a behind-the-scenes consultant who primarily works with other doctors to help them to make diagnoses by means of medical imaging—including x-rays, ultrasound, CT, magnetic resonance imaging (MRI), and similar technologies. Radiologists generally are not the ones who first meet the patient, and rarely hear directly from patients regarding the problems or symptoms they are experiencing.

Sometimes a diagnosis seems clear to physicians when they hear a patient's story and examine him or her, check a blood chemistry value, and so forth, but often the diagnosis remains uncertain. We can infer that physicians turn to medical imaging when there is sufficient diagnostic uncertainty for them to decide between diagnostic possibilities and treatment options (or to decide whether any treatment is even appropriate) for their patient. In many cases, an imaging study can serve to establish the diagnosis that the treating physician suspects, such as when a patient who presents with fever and new onset of right lower quadrant abdominal pain and tenderness to touch is thought to have acute appendicitis, and this diagnosis is later confirmed on a CT scan. In such cases imaging just adds to the doctor's confidence in his or her diagnosis. But sometimes the level of uncertainty is so great that it cannot truly be resolved, and in those cases the doctor must rely on *probabilities* in deciding what to do. If the uncertainty is *too* great, it can be paralyzing, especially when the treatment options are themselves risky, painful, or expensive (or all of the above). The goal of the medical imaging process in such uncertain situations, and truly the most important role of the radiologist, is to try to *reduce* the level of uncertainty present such that the doctor becomes sufficiently comfortable to act—and we can suppose that the radiology process has succeeded once that goal is accomplished. But sometimes the most likely diagnosis—the product of the doctor's examination and the radiologist's best interpretation of the scan or the x-ray—turns out to be wrong. In this chapter we will discuss what it might mean when the final diagnostic outcome does not match the radiology result, which is found in the final radiologist's impression—a question that inevitably brings us back to the question of what constitutes an error versus the high level of uncertainty that the process begins with.

THE PRETEST PROBABILITY OF DISEASE

Bayes Theorem provides a very useful concept in understanding how to interpret any sort of medical test, namely the "pretest probability of disease." In a way, this idea is just common sense: We ask ourselves, what is plausible? Is a four-year-old girl with a cough, for example, likely to have lung cancer? Her pretest probability of that particular disease is extraordinarily low, since lung cancer almost always occurs in older people and typically only after years of smoking. But what if her chest x-ray were to show a spot that looked a lot like a lung cancer? What then? Well, with such a low pretest probability of disease, most radiologists would seek other explanations for the chest x-ray finding.

Getting back to the example of the patient with acute appendicitis: Imagine a young patient with a history of acute-onset, progressively focal pain and tenderness in the right lower abdominal quadrant, with symptoms having evolved over the prior 24 to 48 hours and accompanied by fever. Such a patient would have a very high pretest probability of having acute appendicitis. Some surgeons might be satisfied with the level of certainty (low degree of uncertainty) that remains in the diagnosis after obtaining this patient history and classic physical examination findings, and would thus be willing to take the patient directly to the operating room (OR) to perform an appendectomy. Other surgeons may be less sanguine and will request some confirmatory test, usually CT or MRI, to try to confirm the diagnosis first, before taking the patient to the OR. In such a case the radiologist's interpretive conclusion might be strongly influenced by knowing the clinical history and physical findings.

Now imagine a situation where the CT scan findings aren't absolutely certain, so the radiologist reports back an equivocal imaging result to the surgeon. What happens then? Well, the surgeon may *still* elect to perform immediate surgery based on the high pretest probability that acute appendicitis is present, or he or she may instead choose to "watch and wait," in the event that the final diagnosis "declares itself," leading to surgery later, or perhaps alternatively if the symptoms turn out to be self-limiting and resolve spontaneously, thus allowing surgery to be avoided.

In this way we can begin to think about how the pretest probability of disease, which is influenced by a wide range of factors (age, symptoms, smoking history, family history, etc.), combines with the results of tests—including the diagnostic impressions of radiologists who interpret imaging studies—to lead to a decision. This is a type of reasoning that most physicians use and is how the process of radiology diagnosis usually works.

As a result of years of training, focused study, and supervised experience, radiologists know a great deal about the various ways that disease processes can appear on radiographs and scans, and they also know a lot about the pretest probabilities of a wide range of diseases for patients of varying ages and presentations. In practice, however, radiologists are almost never given complete information as to the patient's (often highly uncertain) clinical presentation, which prompted the doctor to obtain an imaging study in the first place, so radiologists generally must operate with incomplete information as they consider the meaning of the findings they see on the images, including key elements of the history and symptoms that might have changed their ideas about what the "pretest probability of disease" might actually be for any given patient.

LAYERS OF UNCERTAINTY

There is also a great deal of variability inherent in the imaging process itself, and this uncertainty also impacts the radiologist's final impression. First, the images from which radiologists extract information are acquired using a wide range of equipment types and physical techniques, and in turn these images are electronically processed and reconstructed digitally in myriad different ways that vary by equipment vendor, machine settings, patient positioning, and even technologist preferences. These images contain a great deal of information about the patient, but they are also limited in many ways. In a very tangible sense, the information contained within the images is "encoded." It exists in the form of often very subtle differences in ultrasonic audio- or radiofrequency signals or x-ray absorption differences in tissues, reflecting alterations from the normal range of anatomy or physiology of the patient's body, which can be highly variable from patient to patient for equivalent medical conditions.

With regard to these characteristics, the range of what would be considered "normal" usually overlaps the even broader range of "abnormal." Radiologists must therefore continually make mental adjustments of their expectations to account for what amounts to a large and unknowable degree of variation that enters the diagnostic imaging process even before the images are ever viewed—including the aforementioned variations in anatomy from patient to patient, the radiographic technique and patient positioning used, image noise and digital image manipulation, patient size and body composition, patient motion, and widely variable effects of injected contrast-enhancing materials that may be observed within and between patients in both health and disease. (In radiological parlance, "contrast materials" are imaging pharmaceuticals that are injected intravenously into patients during their scans with the goal of improving the information content of the images. They most commonly include iodine-containing materials for x-rays and CT and gadolinium-containing contrast materials for MRI.)

This is not to mention a potentially bewildering variety of physical and biological *artifacts* (strange quirks of the imaging process itself), which occur often enough that they can confuse matters even further. Some are common and easy to understand, such as when the x-ray beam or other signal is blocked from reaching the imaging receptor due to the presence of metal in the patient's body. Others are more subtle, with causes that are less clear, and these types of artifacts can be more confusing, and therefore risky.

To top it all off, there is the fact that there are really very few truly pathognomonic appearances of any given disease on imaging. To the contrary, many very distinct disease states can appear essentially identical to each other on scans and x-ray images, or at least share a common imaging appearance; see, for example, Figure 1.1.

So there are indeed a great many factors inherent to the imaging process itself, factors that simply cannot be changed or improved very much in the hope of increasing diagnostic accuracy.

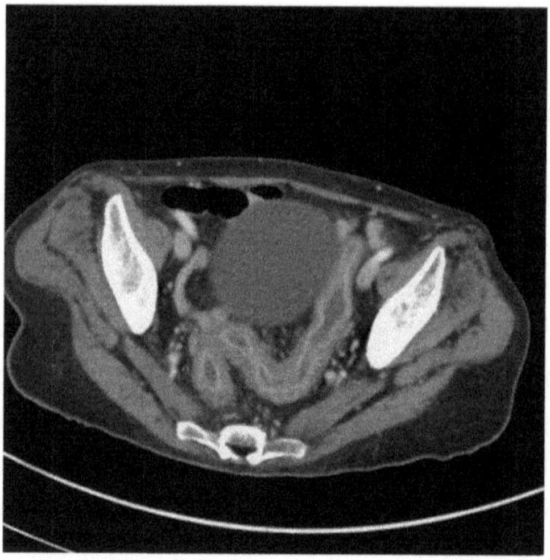

FIGURE 1.1. CT image of a patient showing a region of abnormal-appearing bowel. The appearance is not pathognomonic for any particular disease; a reasonable differential diagnosis for this CT appearance includes ischemia, infection, and inflammatory bowel disease. Although they have a common appearance on imaging, these three distinct disease entities have little in common with each other in terms of etiology, treatment, or prognosis.

Finally, and most importantly, the practice of diagnostic interpretation by radiologists is itself a fallible, human endeavor, one involving complex perceptual, neurophysiological, and cognitive processes by individual observers under a wide range of circumstances and settings, and influenced by a range of nonquantifiable variables, such as fatigue, the observer's level of alertness, inadvertent mind-wandering, mental or cognitive biases, the observer's state of well-being, extrinsic pressures on him or her, interruptions, and the like. These "human observer" factors may ultimately be what establish the absolute limits to the overall accuracy and diagnostic performance of the imaging process, so we will focus on these topics in subsequent chapters.

Some experts, particularly in the computer sciences, have suggested that the human observer may ultimately (and perhaps soon) be replaced by computer algorithms, a topic we will address in the final chapter of this book. This idea has become an area of intense interest in the field and has generated considerable discussion at radiology national meetings and in online discussion forums. Most experts agree that, given the complexity of the radiologist's task and the current state of the art in artificial intelligence, the human observer will probably not be eliminated, at least not completely, any time soon. Thus I remain convinced that the issue of improving observer performance and reducing human error is still a worthy and important topic for study.

In this first chapter we will focus on the very high level of uncertainty inherent to the process of medical imaging (including radiological interpretation) itself versus what might be considered a radiological error to highlight the considerable degree of overlap that exists between the two. What is the boundary between individual observer error and the fundamental uncertainties inherent to the processes? This basic understanding of the overlap of uncertainty and error will help to inform our later discussion of research on error reduction and efforts aimed at preventing patient harms as a result of diagnostic error.

As noted earlier, the radiologist's task begins with visual detection of (often very subtle) variations in electronically displayed shades of gray[1] in an image (or set of images), which form the basis of discrimination. These visually detectable grayscale variations are the "findings" from which the presence or absence of disease is inferred. The radiologist must visually identify the ones that are meaningful, ignore those that are not, and ultimately decide which of the innumerable variations of grayscale intensity might have diagnostic significance for a particular patient. In other words, some of this is "nature," abilities that are built in to the human visual system, and some of it is "nurture," related to the specific expertise that the interpreting radiologist has gained from years of training and experience. With regard to the neuropsychological aspects of the task, for example, we know from cognitive/visual research studies that the likelihood of missing an abnormality on any given imaging examination is increased when findings of that type are rare (low prevalence), although the reasons for this are not known. The most common type of error, simply not seeing something that may be quite obvious in retrospect, occurs randomly and with a low—but stubborn—frequency.

Cognitive biases are known to affect *what radiologists see* as well as the conclusions they reach. Many of these biases are acquired during the radiologists' lengthy residency and (often) fellowship training, which may span six or seven years after medical school, an intensive period of supervised work and focused study in which radiologists-in-training will likely have viewed more than 60,000 medical imaging studies under the tutelage of more senior radiologists. With regard to biases in radiologists' training, we know that radiologists are strongly influenced by their teachers' and mentors' beliefs and habits of practice. Despite decades-long efforts aimed at standardizing the radiology residency curriculum and national certification testing, it is inevitable that individual radiologists' training may or may not have included certain topics and that their understanding of the importance of particular appearances seen on images will be variably influenced by the opinions held by the senior radiologists who trained them.

Further along these lines, each radiologist will inevitably bring into the interpretive process his or her own variable opinions and biases as to the likelihood of various disease entities being present in the population of patients he or she is serving. Based on this, a radiologist's diagnoses may be expected to be more accurate in his or her "home" setting, where he or she is more familiar with the health issues relevant to the population, than if the same radiologist provides services in another, different setting, where the population health issues differ. One can imagine this might be the case if radiologists from an urban, academic medical center provide tele-radiology services and thus come to interpret studies from a faraway rural hospital center delivered to them electronically across the miles.

Later we will delve into the relevant science of visual image perception as it relates to diagnostic radiology; for our purposes here it is worth mentioning that such a complex process as visual image interpretation is inevitably subject to a wide degree of variation—a fact that numerous studies have confirmed exists, both between radiologists and even "within" individual radiologists, leading to the same imaging inputs sometimes producing very different outputs in terms of final diagnostic impressions, just as one might expect. And there is reason to believe that the observer's cognition and his or her perception run on a two-way street: each affects the other, often in unpredictable or counterintuitive ways. Thus what is "seen" on an image sometimes varies considerably based on the observer's expectations and mental state, just as perceived features on an image may influence the physician's reasoning.

Finally, in terms of the interpretive process it must be stressed that in the real world, radiologists, like other physicians, rarely have the luxury of time; instead, they are generally performing their duties under considerable time pressure due to the urgency of the clinical questions they must address, the number of patients needing care, and so forth. When inevitably faced with uncertainty, the radiologist has precious little time to reflect or remain indecisive. In the world of medicine, dithering or inaction is rarely an option, despite the high degree of uncertainty that prevails. The radiologist usually must make a decision on each case fairly quickly and offer his or her best explanation of the findings despite incomplete information—as do essentially all physicians across all medical and surgical specialties.

Considering all of this, it should not be surprising that both the process and the performance outcomes of radiological diagnoses are subject to a great deal of variability. For example, in a 2010 study performed at the Massachusetts General Hospital by Abujudeh et al., a number of leading subspecialist radiologists reinterpreted a series of abdominal/pelvic CT studies that had been previously interpreted by members of their elite group, including cases they themselves had interpreted as well as those interpreted by one or another of their close colleagues. In these "second-look" image reinterpretations, which were performed blinded to patient outcomes, the members of this expert radiologist panel were found to disagree with each other on more than 30% of the CT studies; they also disagreed with *their own original readings more than 25% of the time!* The results are similar in published studies of mammography double-readings.

Strictly speaking, the existence of inter- and intra-observer variation of this large a magnitude is not in itself proof that one or another radiologist has made an error, or even that radiologists as a group are particularly error-prone. To the contrary, I believe that these results merely illustrate the degree to which high levels of uncertainty and inherent variability in the process of radiological diagnosis establishes the boundaries in which imaging tests can ultimately be relied upon, even in the hands of the most qualified and distinguished of experts.

There is also a subtle semantic problem that I believe contributes to misunderstanding about radiological interpretation, especially as regards the relative contribution of error as opposed to uncertainty in radiological practice—and that is the common use of the word "reading" to describe the radiologist's task. We commonly refer to the "reading" of a radiological imaging test or scan, as if the answer were there in black and white, as clear as the words on this page, and all the radiologist needs to do is to open the case and "read" it. In reality, the process is much more like solving a puzzle, one where the pieces aren't always even readily apparent and don't fit together perfectly well, and in many cases where it is not even clear to the observer that there is even a puzzle there to solve.

TEST PERFORMANCE AND THE ROC CURVE

The inherent limitations of any particular imaging test can be illustrated in a graphic known as the "receiver-operator characteristic," or ROC curve, which is a statistical graphic tool derived from signal detection theory in which the true-positive rate of the test in question (i.e., with the "true positive" rate being the fraction of the time that a positive test result corresponds to the patient actually having the disease) is graphed against the same test's false-positive rate (i.e., the fraction of the time that a positive test result occurs even though the patient does not have the disease). An example of an ROC curve is shown in Figure 1.2. The ROC curve was invented during World War II as a tool to evaluate the reliability of detecting enemy objects in battle

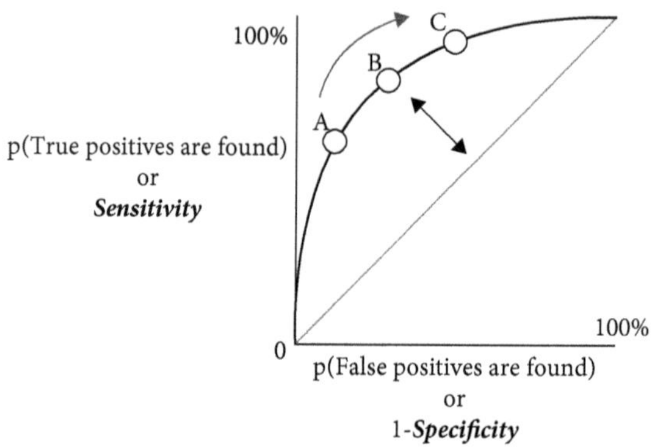

FIGURE 1.2. Sample ROC curve showing a test with fairly good discriminating performance. In an ROC curve, true positives are plotted against false positives (which is mathematically equivalent to the sensitivity of the test vs. 1 – the test's specificity). The diagonal line emanating from the origin would indicate a test with zero discriminating power, whereas the curved line, more typical for a diagnostic imaging test, reflects a study capable of discriminating between the two diagnostic extremes (disease present vs. disease not present) with at least a fair degree of reliability. Points A, B, and C on the curve reflect varying thresholds of detection, thresholds that are generally determined by the observer. The curved arrow indicates a rising threshold. Note that at point A there are fewer false positives but also fewer true positives; at point C there are both more true positives and false positives, illustrating the inherent tradeoffs involved between sensitivity and specificity for any given test. Image credit: Masato (http://creativecommons.org/licenses/by-sa/3.0)], via Wikimedia Commons

by use of radar (a new invention at the time), but the technique was readily adopted in radiology research as a useful way to evaluate which imaging tests are capable of discriminating between various diagnostic possibilities. Although a full discussion of ROC curves and their uses is beyond our scope here, the interested reader is encouraged to explore the suggested reading list at the end of this chapter.

ROC curves have become a standard, classic tool in radiology research, one that by its very nature acknowledges that the performance of any diagnostic test is imperfect. It is understood by users of ROC curves that test performance is inevitably influenced by a range of factors that cannot themselves be separately measured, including unknowable technical, clinical, and human observer factors. The ROC curve is thus an aggregate measurement used to assess the entire imaging process under evaluation, essentially from the point of patient selection to final image interpretation, bundling together all of the inherent uncertainties of each step in the process, whether they are individually and separately known or unknown. It can be used to easily identify tests that have a strong discriminating power from those that do not, as Figure 1.2 shows. It also can be used to graphically illustrate the idea that the observer can, in a very real sense, adjust the thresholds of discrimination—such that any given test may be used for different purposes—and illustrates the inherent tradeoffs involved. For example, it may be desirable for the "action threshold" of a test to be low (i.e., operating closer to the left side of the curve, such as in point A of the illustration), where sensitivity approaches 100% but specificity is low (not so many cases are missed, but at the cost of having lots of false alarms) versus setting the threshold of a test to be high (i.e., operating closer to the right-hand side of the curve, such as in point C on the illustration), where sensitivity is lower and so some cases will be missed, but in exchange for this the bulk of the cases that test positive are going to be "true." The beauty of the ROC curve is that it provides a graphic illustration of the situation that radiologists find themselves in—trying to find the "sweet spot" where detection of disease is optimized and false alarms are minimized, as in point B. But I should emphasize that point A is not necessarily worse or better than points B or C; rather, one or the other may be *more or less appropriate* depending on the goals of the test.

RADIOLOGIST PERFORMANCE MEASURES: OBSERVER "ERROR" VERSUS RADIOLOGICAL UNCERTAINTY

In considering the fundamental question as to when uncertainty ends and interpreter error begins, there are few guideposts. To the extent that a defining reference standard may exist to constitute a "correct" diagnostic interpretation, such as an autopsy or biopsy result, or perhaps some objective discovery made at the time of surgery that serves to establish the final diagnosis, we can potentially define a radiological error as having occurred when a radiologist's diagnostic impression differs substantively from such a reference standard.

However, these sorts of "gold standards" are rarely, if ever, available in practice. In fact, their *own* reliability has not been well established. A handful of studies undertaken over recent decades have shown that considerable variability also exists between pathologists' final diagnoses from surgical specimens and at autopsy when different pathologists reexamine the same specimens—data that raise the issue of the limits of reliability of such traditional reference standards as the "final" pathological diagnosis or even the autopsy as an absolute or objective final standard against which to measure clinical and radiological "diagnostic error." While a number of studies have been undertaken to estimate the size of radiological interpretive error, and a great many published studies compare the diagnostic performance of all physicians to a "final diagnosis" at autopsy, there has actually been very little scholarship (and no well-controlled studies have been published to my knowledge) that gives a measurement or even an estimate of the inherent error rate of autopsies themselves. But in a recently published study of pathologists interpreting single standardized breast biopsy specimens to evaluate their reliability in the diagnosis of breast cancer, performed by Dr. Joanne Elmore at the University of Washington in Seattle, there was only 75.3% agreement between the interpretations of individual pathologists and the reference standard of expert consensus—which implies the inter-interpreter reliability of expert pathologists is actually remarkably similar to that of expert radiologists. In fact, the approximate 75% figure almost exactly matches prior results published by Dr. Elmore's group back in 1993 when they evaluated inter-reader agreement of radiologists interpreting screening mammograms. This level of concordance should perhaps not be surprising considering that very similar training, expertise, and visual-perceptual processes are involved. This concordance also provides hope for those of us who want to improve observer performance: it seems likely that anything that can improve the performance of one class of visual diagnosticians (e.g., radiologists) would also be effective for other classes of clinical observers (e.g., pathologists, dermatologists, endoscopists).

PEER REVIEW AS A REFERENCE STANDARD FOR ERROR

In practice, we radiologists have more commonly defined what constitutes an "error" for ourselves without need for reference to a gold standard. In radiology, an interpretive error is generally assumed to have occurred when a radiologist's interpretation of an imaging study deviates from a *group consensus of peer radiologists* with similar credentials and expertise. The standard applied is generally what the group consensus feels that "most" radiologists under "usual circumstances" would have concluded regarding the findings of the imaging study under question, which is presumed to include an understanding of the variability of the finding, the likelihood of the diagnosis in the particular patient population being studied, and the limitations of the technology. This was essentially the same standard used in the study of pathologists' inter-observer agreement noted above. This approach is also similar to the most common standard applied in medical malpractice litigation cases by radiology expert witnesses, except that in the case of malpractice litigation the group consensus is substituted with the expert opinion of a single observer—typically a paid expert witness. In legal proceedings we often note that the expert retained by the plaintiff's side has an opinion that is diametrically opposed to that of the expert retained by the defense side, requiring the jury to decide which of the "dueling experts" is more credible, but that is a different matter; we will explore the medicolegal response to error in greater depth in Chapter 3.

Typically in radiology peer review, a standing committee or ad hoc peer group consensus is generally reached in *retrospect* once the patient's final diagnosis and outcome is known, and thus it really does not fully take the degree of *prospective* uncertainty into account, although in practice the reviewers generally make an honest effort to do so. Any outlying or dissenting opinions, when voiced, are typically overruled by majority vote in peer review committees. So, in essence, in the peer review process the validity of a single expert radiologist's *individual* opinion is deemed to be subjective, and the accuracy of it is ultimately judged by the standard of . . . *group* opinion! Perhaps it is best to classify the supposed errors identified in this process as *discrepant diagnoses*, since it is not possible to truly know whether the reviewers or the reviewed radiologist is ultimately correct.

While such group consensus may be the best measure of radiologist accuracy available, the problems with this approach are, I think, are clearly evident, the principal one being the embedded cultural biases of the reviewers. Perhaps surprisingly considering human nature, most peer review committees appear to judge the majority of missed radiological findings somewhat harshly as errors, and in doing so are likely biased by hindsight, once the results of subsequent studies or the ultimate patient outcomes are known. In the practice of peer review in radiology I have observed very little in the way of error minimization over the years, and generally no attempt to "whitewash" or falsely elevate measures of performance accuracy occur, but rather I have frequently observed a very strong tendency to the contrary. Part of this stems from a culture where we all expect to "take our lumps" in peer review, and partly this is due to the simple fact that it is very difficult to "un-see" a finding once it is pointed out—and it is very difficult to judge how likely a finding would have been seen in prospect.

Consider the logo for Federal Express (Fig. 1.3). Most people are unaware that it includes a rightward-pointing arrow between the E and the x—but once this previously "hidden" feature is pointed out, it cannot be "un-seen." Peer review could be made much more accurate by use of completely blinded prospective double-readings of studies, but this approach is rarely used due to the cost and workload involved.

Almost every radiology practice in the United States today uses peer review as a quality improvement measure; most use the basic approach created by the American College of Radiology (ACR)'s RadPeer™ system, in which a small sample of the hospital's cases, typically on the order of 5%, are randomly selected for review once a new or follow-up study has been completed. The radiologist who is interpreting the new follow-up study is tasked to evaluate the

FIGURE 1.3. The FedEx corporate logo. Note the arrow between the E and the x.

original interpretation of the prior study to which he or she is making a comparison in the light of the newer study he or she is currently interpreting. In my own medical center, as in many others, the randomized selection and "presentation" of cases for peer review is done by an automated system, the scores are submitted electronically, and the cases flagged as having significant disagreement are reviewed later by a committee. The ACR maintains a large database in which the scores from these diagnostic discrepancies are aggregated and later are used in research. The system works reasonably well as a measure of overall system performance and can occasionally identify areas of concern with individual observer performance; however, it is a rather poor overall error detector, since typically 95% or more cases are never reviewed. Thus, peer review is a very poor "failsafe" or trigger tool for error detection in radiology.

Scoring under the system is based on the reviewers' consensus of how conspicuous a finding is, and how likely the reviewer feels the "average" radiologist practicing under "usual" circumstances would have reached the same or a different conclusion than the original interpreting radiologist did—an amazingly difficult task in reality. It thus tends to score perceptual errors (i.e., those in which a finding is missed but is fairly readily detectable in retrospect) more harshly, whereas more subtle findings that are missed are often down-scored in this system; in doing so, the RadPeer scoring system attempts to distinguish between what might be considered an "understandable miss" versus an "error," with the presumption that any corrective measures would be different between these discrepancy types. (Mercifully, a former category of a super-highly obvious missed finding, which could be considered to be an "egregious miss," has recently been eliminated from the scoring scheme!)

Errors are also scored as to their likelihood of having caused any clinical harm or impact to the individual patient involved, which is often determined in hindsight after subsequent imaging.

There is an unstated fatalism in the idea that some findings are too subtle to be reliably detected, and an unstated overconfidence, I feel, to suggest that other types of errors might be eliminated if only radiologists were "more careful." Neither conclusion appears to be supported by evidence, although it remains a too-common practice in centers throughout the United States for individual radiologists to be counseled, or even threatened with dismissal from their jobs, based on peer review scores.

The ACR's RadPeer™ Committee, a group of radiologists charged with developing and improving the ACR's peer review system and database, and also with guiding radiology practices in its best use (a committee on which I served a four-year term), has over the years published a small number of policy "white papers" advising radiology practices on how best to use the ACR's peer review model. In these publications the committee has repeatedly noted, for example, that the RadPeer scores and data must be used only for *quality improvement* and always in a *nonpunitive fashion* rather than as a tool primarily to identify and remediate (or remove) underperforming individuals. This is because it has long been established that fear of recrimination—whether for oneself or for one's colleagues—will inevitably have the effect of reducing the reporting of errors and discrepant findings. This understanding makes up the essence of the call for a "blameless culture" in the identification of errors in health-care organizations.

In a recent white paper on the use of RadPeer, the committee went on to specifically warn against including peer review data on radiologists' individual performance reviews for hospital privileges or other factors directly impacting their employment. Yet the practice of incorporating peer review scores into individual practitioners' Ongoing Professional Performance Evaluations, twice-yearly evaluations mandated by the Joint Commission for Hospital Accreditation, remains widespread. Accordingly, the data from peer review are likely skewed toward underdetection with regards to error detection, in addition to their being contaminated, as we noted above, from hindsight bias. As a result, peer review data are a very poor reference standard for radiologist error, and sadly they shed little or no light on the underlying issue of the relationship between what constitutes "error" and what merely reflects the high level of inherent uncertainty in the process itself.

SCREENING EXAMINATIONS AND ERROR

Screening examinations—the class of radiology tests where a large asymptomatic population (and thus a relatively low level of pretest probability of disease) is examined in the hope of detecting a few cases of disease at an earlier stage—are a special case with regard to uncertainty and error. In the screening scenario, the prevalence of true-positive findings is low and the false alarm rate is high. Consider the well-documented 15% failure rate of screening mammography and the high rate of false positives, which is reported to exceed 60%. Recall that most patients undergo annual screening mammography in the absence of disease, with only a very small fraction of true-positive screening mammograms leading to a diagnosis of breast cancer—hopefully earlier than could be attained using other methods, such as annual physician breast exams or monthly patient breast self-exams. But for most women of screening age, breast cancers grow fairly slowly in the early stages while they are small, and often a cancer is present for years before it can be detected. As a result, virtually every "positive" mammogram, on which a lesion is detected, is thus preceded by at least one prior mammogram, obtained one year earlier, on which no lesion was detected but in retrospect the presence of the cancer can be identified. Clearly this reality does not point to what would amount to an essentially 100% error rate of screening mammography, but instead mammography serves as a small test laboratory for the overall problem of image-based diagnosis, the reality of uncertainty, and the question of what constitutes error for radiology as a whole. As experience grows in other types of image-based screening, such as the use of annual low-dose chest CT scans in smokers for lung cancer screening, the experience will likely be similar.

RADIOLOGY ERROR VERSUS UNCERTAINTY: SOME CONCLUSIONS

While it is difficult to draw a line or calculate a fraction, it seems clear on deeper analysis of radiologists' errors in practice that a significant portion of discrepant diagnoses in radiology—those incorrect diagnoses typically discovered by retrospective peer review after the final diagnosis has become clearer with time, are actually more likely to represent manifestations of the underlying uncertainty inherent to the diagnostic imaging process. Often detection of an earlier "error" is made possible because the underlying disease progresses over time and ultimately "declares itself," becoming more obvious on subsequent imaging studies used for comparison. This realization has far-reaching ramifications in the design of future research to help understand the causes of error as well as efforts to develop strategies for error reduction and harms mitigation that have a chance of working, as well as the general realm of individual accountability for error, an area that has legal, regulatory, and cultural aspects in the current practice of medicine and is a topic to which we shall return later.

NOTE

1. Although the human eye is capable of discriminating approximately 500 shades of gray, digital images are typically constructed with 8 bits per pixel, which yields 256 shades of gray. Digital mammograms may use many more

FURTHER READING

Abujudeh HH, Boland GW, Kaewlai R, et al. Abdominal and pelvic computed tomography (CT) interpretation: discrepancy rates among experienced radiologists. *Eur Radiol* 2010; 20(8):1952–1957.

Berlin L. Accuracy of diagnostic procedures: has it improved over the past five decades? *Am J Roentgenol* 2007; 188(5):1173–1178.

Berlin L. Radiologic errors, past, present and future. *Diagnosis* 2014; 1(1):79–84. doi:https://doi.org/10.1515/dx-2013-0012

Brady AP. Error and discrepancy in radiology: inevitable or avoidable? *Insights Imaging* 2017; 8(1):171–182. PMID 27928712; doi:10.1007/s13244-016-0534-1

Bruno MA, Hollenbeak CS. Statistical tools and methods, Chapter 18 in Abujudeh HH, Bruno MA, eds. *Radiology Noninterpretive Skills—The Requisites.* 1st ed. Philadelphia: Elsevier Scientific Press, 2017. ISBN: 9780323462976

Bruno MA, Walker EA, Abujudeh HH. Understanding and confronting our mistakes: the epidemiology of error in radiology and strategies for error reduction. *RadioGraphics* 2015; 35:1668–1676. https://doi.org/10.1148/rg.2015150023

Elmore JG, Longton GM, Carney PA, et al. Diagnostic concordance among pathologists interpreting breast biopsy specimens. *JAMA* 2015; 313(11):1122–1132.

Elmore JG, Wells MPH, Lee CH, et al. Variability in radiologists' interpretations of mammograms. *N Engl J Med* 1993; 51(22):1493–1499.

Firestein S. *Failure: Why Science Is So Successful.* New York: Oxford University Press, 2016. ISBN-13: 978-0199390106

Hatch S. *Snowball in a Blizzard: A Physician's Notes on Uncertainty in Medicine.* New York: Basic Books, 2016. ISBN-13: 978-0465050642

Hillman B, Goldsmith J. *The Sorcerer's Apprentice: How Medical Imaging Is Changing Health Care.* New York: Oxford University Press, 2011. ISBN-13: 978-0195386967

2

TAXONOMY AND EPIDEMIOLOGY OF ERROR IN RADIOLOGY

Prevalence and Types of Radiological Error

A 24-year-old woman comes to the emergency department (ED) complaining of fever, chills, and progressive abdominal pain. The pain was generalized and was felt throughout the abdomen when it began, but over the past day and a half she noticed that it had become more localized to the right lower quadrant. The emergency physician also noted that the patient was very tender to touch in that area, so she strongly suspects a diagnosis of acute appendicitis. A computed tomography (CT) scan is requested to confirm the diagnosis. The radiologist who interprets the study notes the history: "Rule out appy." The images show a classic example of acute appendicitis, with a swollen, thickened appendix in just the right location, containing a small calcium stone known as an appendicolith, and surrounded by inflammatory changes. She dictates and electronically signs a report with the conclusion: "Acute appendicitis." Since this was the clinical question posed by the referring doctor, the radiologist concludes that this diagnosis is not a surprise and does not choose to interrupt the ED physician with a telephone call, as she might have if the provided history were different and the diagnosis was surprising. Once signed, the electronic report is immediately available in the patient's electronic medical record, which is checked by the ED physician. Unfortunately, the referring doctor makes a mistake when she goes to check the patient's results: that same patient had a prior scan on the exact same date four years earlier, and the ER physician opens that old report by mistake, not noticing the year in the date of exam. Four years ago, the appendix appeared normal, and at that time the radiologist clearly stated "normal appendix." The ER physician thus erroneously concludes that the scan on her patient showed a normal appendix. She feels momentarily perplexed that the CT findings disagree with her impression from the patient's history and physical examination. She wonders whether it would be worthwhile to ask the radiologist if there may have been some sort of mistake, but ultimately she shrugs off this thought, decides to rely on the scan, and sends the patient home, telling the patient that she was mistaken in her suspicion of appendicitis, since her appendix is normal on CT. A day later, the patient returns to the ED with worsened and persistent symptoms and is examined again, and the doctor's error is discovered. Luckily, no lasting harm resulted. The ED physician is angry with the radiologist for not phoning her with the positive result.

Most error classifications in radiology have focused on two primary error types, those of *perception*, where objects on images are simply not seen, and those of *cognition*, where the findings are visually detected and duly noted, but the meaning of those findings is misinterpreted. But there is reason to believe that this type of thinking about errors may be overly simplistic.

From past studies of the types of errors that radiologists are prone to make, dating from the pioneering work of the radiologist Dr. Leo Henry Garland in 1949, there is reason to believe that most radiological errors are errors of *omission*—false-negative interpretations where some finding on the images, detectable in retrospect, was not fully appreciated at the time of the original interpretation. We know that these types of errors are fairly common, but most of them carry statistically little chance of having a direct or immediate consequence on a patient's final outcome.

We also know that errors of *commission* occur as well—false-positive interpretations where the radiologist's conclusion was not truly supported by the imaging findings. These types of errors appear to be considerably less frequent and may be somewhat more difficult to detect experimentally. The unavoidable conclusion that most radiologist errors go undetected also suggests the urgent need for future research on developing better means to detect radiologists' errors more promptly in order to reduce the risk of patient harm.

The most common type of radiologist errors (by frequency of occurrence) have been shown to be perceptual lapses—that is, errors that occur when a radiologist simply fails to perceive a finding on an image. Sometimes such findings are subtle, but they are often readily apparent in retrospect (as in the example shown in Fig. 2.1). This

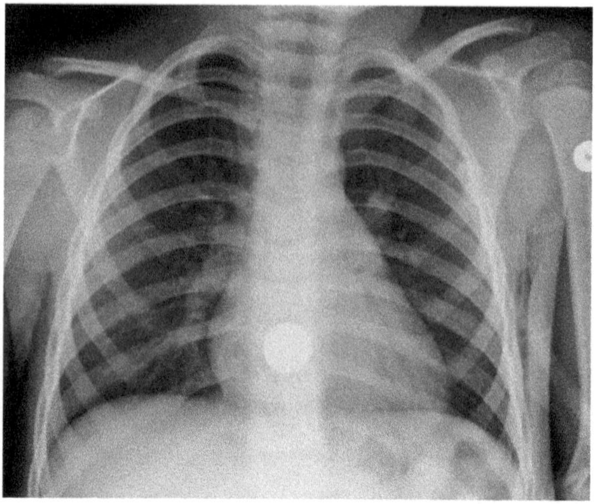

FIGURE 2.1. An example of a perceptual error. Anteroposterior (AP) radiograph of the chest reveals a swallowed coin within the esophagus. This finding was missed—twice—by a skilled and experienced subspecialist radiologist. Reprinted with permission from Bruno et al., *RadioGraphics*, 2015.

phenomenon, which radiologists commonly call a "missed finding," accounts for between 60% and 80% of all radiologist interpretive errors in most series dating back several decades to the time of Dr. Garland's original reports. These errors appear to occur randomly at a more or less fixed occurrence level for all observers. The underlying reasons for this type of error are not fully understood but are believed to be complex and multifactorial, related to some fundamental (biological) aspect of perception and visual-based discrimination tasks in human observers, including cognitive, neurophysiological, attentional, and functional/psychological factors. They are clearly not, as experimental evidence has shown, generally the result of lapses in work ethic or deficient knowledge or skill and almost never are due to poor visual acuity on the part of individual radiologists. This type of error has, thus far, proved intractable to a wide range of intervention strategies.

The second most common type of error in radiology is related to a well-described phenomenon known as "satisfaction of search," which was first described in 1990. Satisfaction of search is an error where a radiologist fails to detect a second abnormality present on an imaging study after finding a first abnormality (and the first is not always the most obvious one), perhaps, as the name implies, because of prematurely ceasing to search the images as a result of detecting a "satisfying" finding. Often the "satisfying" finding is one that explains the patient's clinical symptoms, but the reason why any given finding might be somehow psychologically "satisfying" to the radiologist is unclear.

In the comprehensive study of the frequency of various types of radiological error published by Kim and Mansfield in 2014, this "satisfaction of search" phenomenon was associated with a large number of radiologist interpretive errors, approximately half as many as due to perceptual errors. Combined, these two categories of error dwarfed all other apparent error causes and types in their series, together accounting for 64% of detected errors, an observation which interpretation of bone radiographs, noted that the detection rate for the first abnormal finding in a matches previously published similar studies fairly well. Ashman et al., analyzing errors in radiology cases that contained at least three abnormalities observed a very high detection rate for one of the three (the initial abnormality that was found) but the second and third abnormal findings were each only discovered approximately 40% of the time. Most radiologists will report having succumbed to this type of error at some time during their career, although a more recent study by Krupinski, Berbaum, et al. published in 2010 failed to detect this type of error in their study population, which suggests that its prevalence may possibly have been overestimated.

Other types of radiologist interpretive errors have been described, including by Kim and Mansfield (Table 2.1), that may be more amenable to intervention, such as gaps in the radiologist's knowledge, the presence of cognitive biases, the failure to detect findings located in unexpected places on the images, or simply a misunderstanding of the pretest probability of disease—which in turn may sometimes be due to misleading or erroneous clinical information having been provided to the interpreting radiologist by the patient's doctor. In addition to perceptual errors and the phenomenon of satisfaction-of-search errors, Kim and Mansfield also described 10 additional types/categories of radiologist errors, each accounting for a small fraction of the total number of radiologist errors detected in their study. Technical limitations were found to produce false-negative results in only 2% of cases, presumably all detected when repeat studies were obtained that were technically superior. Kim and Mansfield attributed 2% of the errors in their series to faulty or incomplete clinical history, and 7% of errors were attributed to the location of the abnormality being unusual or outside of the radiologist's area of interest on an image. A total of 9% were attributed to faulty reasoning on the part of the interpreting radiologist, a category that included all of the errors of commission, which these investigators termed "over-reading," including false-positive results and errors where the radiologist's conclusion did not match the final diagnosis. This category of error would be most closely related to the effect of cognitive biases on the part of the interpreting physician. It is worth

Table 2.1

THE KIM AND MANSFIELD CLASSIFICATION SCHEME FOR ERRORS IN DIAGNOSTIC RADIOLOGY

Cause of error	*Explanation*	*Percentage occurred*
Complacency	A finding is appreciated that is believed to be more serious than it actually is (false positive).	0.9%
Faulty Reasoning	A finding is appreciated and interpreted as abnormal but is attributed to the wrong cause (true positive misclassified).	9%, third most common error. See Figure 2.2.
Lack of Knowledge	A finding is seen but is misattributed because of a lack of knowledge on the part of the interpreter.	3%
Under-reading *(the missed finding)*	A finding is present on the image but is missed (perceptual error).	42%, the most common error type. See Figure 2.1.
Poor Communication	An abnormality is identified and interpreted correctly, but the message does not reach the clinician.	0%
Technique	A finding is missed because of the limitations of examination or technique.	2%
Prior Examination	A finding is either not perceived or is misattributed because of failure to consult prior radiological studies or reports.	5%

(*continued*)

Table 2.1 CONTINUED

Cause of error	*Explanation*	*Percentage occurred*
History	A finding is either not perceived or is misatrributed because of inaccurate or incomplete clinical history.	2%
Location	A finding is missed because of the location of a lesion outside the area of interest on an image.	7%, fourth most common error. See Figure 2.3.
Satisfaction of Search	One of two or more abnormalities on an image is missed; historically this was thought to occur because of failure to continue to search for additional abnormalities after the first abnormality was found.	22%, second most common error
Complication	A complication from an invasive procedure	0.5%
Satisfaction of Report	A finding was either not seen or was misattributed to the wrong cause because of bias related to overreliance of the radiology report of the previous examinations.	6%

Based on Kim and Mansfield, *American Journal of Roentgenology*, 2014.

noting that only 3% of errors were attributed to knowledge gaps on the part of the interpreting radiologist.

What was most alarming, however, was Kim and Mansfield's observation in this study that radiologists' errors, from whatever apparent cause, tended to be repeated or propagated on subsequent radiological examinations of the same patient, either by the original radiologists or by other radiologists who interpreted the subsequent studies that perpetuated the error. The cause that underlies the observed repetition or propagation of errors on subsequent radiological examinations remains uninvestigated, although underlying cognitive biases on the part of the involved radiologists are strongly implicated.

ERROR PREVALENCE

A number of published studies have assessed the frequency with which one radiologist's interpretation has disagreed with another's; this provides a low-end estimate of the prevalence of error in the daily practice of radiology. That rate is probably on the order of 3% to 5%. Given that approximately 1 billion radiological studies are obtained and interpreted annually worldwide, and choosing the middle of that error range, 4%, we arrive at a rough estimate of 40 million radiologist interpretive errors each year throughout the world. This does not include the false-positive rate of mammography, which is thought to exceed 60%, and does not include intra-observer variation, as was discussed in Chapter 1. Conversely, of course, this means that radiologists succeed, despite the extreme difficulty of their task, around 96% of the time in routine daily practice, which involves the fairly rapid review of a large number of diagnostic images each day.

RADIOLOGISTS' UNIQUE AND REMARKABLE EXPERTISE

Since the most common radiologist error type is the simple failure to perceive an abnormality that is detectable in retrospect, it would follow that gaining an understanding of the underlying processes involved in image perception and visual detection by radiologists would be of great value. This subject will be explored more deeply in Chapter 4. It is worth emphasizing here that despite our focus thus far on radiologists' lapses, as a group, radiologists by and large possess an extraordinarily high degree of expertise for the detection of abnormalities on medical images, as one might expect given their years of training and the sheer number of medical images they have viewed. Interestingly, this unique expertise seems to be very specific and does not readily translate to other types of tasks, including other visual-perception–based tasks, such as finding Waldo in a "Where's Waldo" puzzle, which radiologists find just as difficult as anyone else. Radiologists are as susceptible as untrained persons to so-called optical illusions on images or missing objects on nonmedical images during visual assessments of "change blindness," for example. Change blindness is a phenomenon where people are not readily able to detect when some aspect of a visual image, usually something in the background of the main subject, has changed—such as when a large mountain appears in the background of a photo of two people. There is also no reason to believe that radiologists' performance on untrained similar visual discrimination tasks, such as discriminating between T's and L's of varying orientations (commonly used by visual psychologists) would be any better than that of laypersons (Fig. 2.2).

To understand what is going on with radiologists, therefore, it is worthwhile to better understand what aspects of the visual data-gathering process are unique to their particular expertise as possible sources of error, as well as understanding those vulnerabilities to making errors that they share with all human observers.

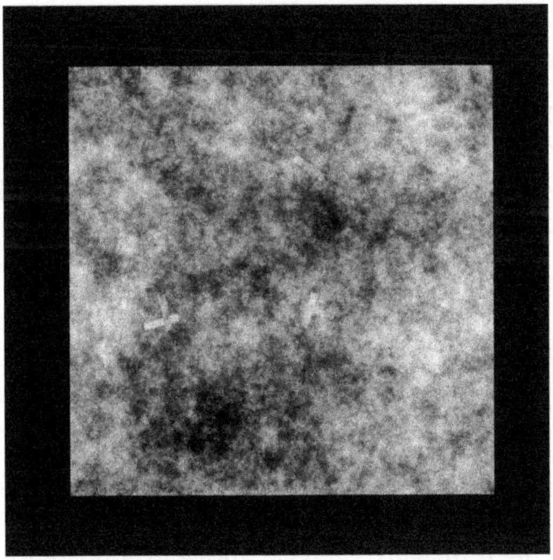

FIGURE 2.2. Psychometric visual test pattern for discriminating the target "T" on a camouflaged background containing "L's."

THE HALF-SECOND GLIMPSE

Experiments on experienced, expert radiologists performed over the past decade have shown a very remarkable result: radiologists are able to detect abnormalities on medical images *extraordinarily quickly*, and often with very little need for systematic searching of the images. Often an abnormality is identified on an image (and the radiologist's eyes move directly to it) in only about half of a second! This is much faster than would be possible if the radiologists had to randomly encounter the abnormality while systematically searching an image, or especially if they needed to think about the finding very much in the process of noticing it. Presumably, there is something happening at an unconscious level based on visual cues in the radiologists' peripheral vision that cause them to direct their gaze—and only subsequently do they become consciously aware of what has been detected. These visual processes and radiologists' eye movements have been studied carefully, but the unconscious aspects are still not well understood; they appear to represent a neurophysiological effect of the radiologists' years of training and experience.

We also have learned from years of careful research that when the "half-second detection" does not succeed and radiologists visually search medical images, they often follow a distinct, reproducible pattern of eye movements. Generally, these eye movements do not cover the entire image in the way a device might scan an image (e.g., starting at the top-left corner and finishing at the bottom-right corner); instead, the eyes move around the image in particular ways, looking at some areas and ignoring others. Clearly these search patterns are selective, and we believe that they can be influenced by the radiologist's individual understanding of the clinical question posed or by the pretest probability of disease or injury. What is expected thus influences what is seen. While this may seem perilous, in practice it is remarkably successful; it has been shown that radiologists' accuracy can be as high as 96% for detection and characterization of abnormalities in routine practice, which is many times more accurate than non-radiologists viewing the same images. A notable comparison would be to the performance of airport screeners, for example, who use x-ray equipment for visual detection of weapons or other contraband in luggage: they have a reported overall accuracy of only around 5%.

Since most visual acuity is in the central (foveal) vision, finding an abnormality by visual search generally requires moving the eyes around an image in order to aim the central visual field onto each of many areas of interest, one after another. Often these eye movements are performed in a more or less habitual way by each radiologist, guided by such factors as personal habit, training, and experience, sometimes influenced by the clinical question posed by the referring physician or other heightened pretest suspicion of disease, or even the radiologist's own clinical knowledge of the range of abnormalities being searched for, as mentioned earlier.

Patterns of radiologists' eye movements has been studied for years using eye-tracking technologies. The pattern is generally believed to be fairly consistent for each radiologist and has become known as the radiologist's "search pattern." In these types of studies, specialized equipment is worn that detects the location of the viewer's gaze on a target image. The equipment can delineate the viewer's visual path taken in scanning the image and also can measure the amount of time spent in each area on the image. Studies of this type have found that a radiologist's "dwell time" in certain areas of an image is often related to the ultimate detection of abnormalities in those areas. Interestingly, missed findings are *also* found in areas with relatively longer visual dwell times. In fact, it has been shown that the radiologist's gaze may dwell on, or frequently return to, areas of an image that contain an abnormality that is ultimately missed. This raises the very interesting idea that there seems to be detection of abnormalities on images occurring below the level of conscious awareness and that sometimes this subconscious lesion detection never reaches an "awareness level." Alternatively, this could reflect instances where the radiologist detects and is conscious of an abnormality but the finding is not stored sufficiently in working memory to contribute to the final diagnostic conclusion for a case. This type of finding highlights an important neurocognitive area of vulnerability for radiologist error.

Perhaps the best example of this is in a 2014 study by Mallett et al. in which the dwell time of a radiologist's gaze fixation in certain areas of a radiographic image was measured and was found to strongly relate to the presence of abnormalities in those areas, whether or not the abnormality was consciously detected. In fact, it was observed that the radiologists more frequently returned to areas of an image that contained (consciously undetected) abnormalities!

Poorly understood attributes of attentional focus and working memory may also play an important role in these perceptual errors. One can hypothesize a mechanism of error in which abnormalities might be detected visually by the observer, occurring somehow below the level of consciousness, but are somehow "rejected" and thus not stored in working memory sufficiently to be included in the radiologist's final impression.

Our understanding of how radiologists interpret images continues to evolve from these sorts of studies of the psychology of visual perception. Classically, the radiologists' process was conceived of in the sequence of "search and detect, recognize, and decide," in which the perception of findings is followed an effortful search, in turn followed by a conscious, cognitive decision-making process. That is indeed how we teach radiology to our trainees. But these newer experimental data have upturned this model, and we now understand that some sort of initial global impression, occurring very rapidly and sometimes without conscious awareness, initiates the perception of abnormalities on medical images. This initial "flash perception" then directs any subsequent formal search that may be undertaken. This new understanding can be summarized as "detect and recognize, search, and decide," relegating the search process to a later stage in perception than was considered before. We know now that, quite often, it is only after the most conspicuous or significant abnormalities have already been identified that a more formal search-and-identify process takes place. The change between these approaches appears to occur organically as a result of years of training and experience.

Another issue of immediate clinical importance, especially for those who manage radiology practices, is understanding what the effects of viewer fatigue and interruption to the search pattern might have on the ultimate performance and accuracy of radiologists. Working long hours and having frequent interruptions are common features of modern radiologists' practice, after all, but are potentially amenable to management intervention. Eye-tracking studies done on radiologists have shown that their search pattern can "deteriorate" when they are fatigued, distracted, or interrupted, which presumably can lead to a higher error rate. This research also suggests that the monitoring of eye movements could potentially provide an avenue to reveal otherwise difficult-to-detect perceptual errors, or at least reveal when there is heightened risk of such errors. Fatigue has been more thoroughly studied than the effects of interruption, but interruption does appear to disrupt the normal search patterns, is known to interfere with working memory, and is thus very likely to contribute to the problem of missed findings on images; this may be attributable to loss of attentional focus or, in the case of fatigue, perhaps even on radiologists' ability to maintain attentional focus. Fatigue and distractions could perhaps also lead to a higher degree of observer "mind-wandering," the human tendency for internal self-distraction from thoughts generated "internally" (i.e., within the mind of the observer), taking his or her attention away from the "external" task of image interpretation. This would also reasonably be expected to affect radiologists' detection performance.

COGNITIVE FACTORS, BIAS, AND "DE-BIASING"

As we consider the complex interplay between radiologists' mental state and the things they see and do not see on images, it may be a good time to consider the effect of cognitive factors, such as radiologists' own biases, on the visual detection process. We know that a great many diagnosis errors, and not just those limited to radiologists, appear to involve faulty or biased cognitive processes—lapses of logic, generally unconscious and emotionally driven, that interfere with clear, objective thinking—but that can actually be adaptive under certain circumstances (including, presumably, those under which human beings evolved). There is a large and growing psychology literature on this subject of cognitive biases: at least 40 clearly delineated cognitive and affective biases have been described, many of which come to bear on this particular issue. It is worth noting, once again, that errors that appear to be entirely perceptual in nature may well have their origin in underlying cognitive biases, leading to visible abnormalities on images being unconsciously overlooked by the viewer, perhaps because of underlying bias in the radiologist's expectations of which findings are likely to be encountered or what types of findings will be searched for within an imaging study.

It is well understood that a radiologist's approach to interpreting a study can be influenced, sometimes unduly so, by the clinical information and context provided. Thus, a referring clinician's provided history and the clinical question posed can influence what radiologists see—or what they think about what they are seeing—either for good or for ill (e.g., if the clinical information that is provided is incorrect or misleading). This phenomenon is known as *framing bias* and is well described in the radiology literature. Many teaching radiologists advise their trainees to examine images and form their opinions before reading the provided history, and then again afterward, in an attempt to minimize this effect.

Biases exist because they help us overcome particular difficulties that we face in life: we are often confronted with too much information and need to simplify our thinking in order to make decisions; we often need to act fast, with incomplete information; we need to assess whether we are facing dangers, or whether the people we encounter will be helpful or potentially harmful to us; we need to sort our myriad daily experiences and decide which ones to remember for the future. Each of the 40 or so described cognitive biases can be thought of as addressing one or more of these issues, and they often act to make our lives easier under certain circumstances. But they often cause problems for us in some aspects of our lives as well. And it is clear that some of them commonly contribute to diagnostic error.

Of the many described cognitive biases, there are three that are felt to have the largest impact on diagnostic accuracy, both in radiology and in other specialties of medicine as well:

- *Anchoring bias*, in which one's initial conclusion is never called into question by the discovery of contrary information; rather, the new information is either discounted or modified to match the prevailing hypothesis. This mental process overlaps with
- *Confirmation bias*, the tendency to search for confirmatory evidence to support one's hypothesis and to discount or ignore contrary evidence, and
- *Availability bias*, in which a diagnosis is considered more likely when one has recently seen a case, especially if the diagnosis was missed in the prior instance.

Accordingly, strategies for "cognitive de-biasing" and related metacognitive interventions (i.e., thinking about thinking) have been advocated to remediate these common types of radiologist error, especially for the most common of the known cognitive biases that may impact clinical reasoning and information-gathering.

The idea of recognizing when a cognitive bias is present and taking active steps to correct for it (termed "cognitive de-biasing") has existed for decades; the idea was well described in 1994 by Wilson and Brekke, who termed it "mental correction." Unfortunately, extensive experimental efforts in recent years to reduce diagnostic error rates in medical specialties outside of radiology by application of de-biasing methodologies and algorithms have been mostly unsuccessful. For example, in a recent meta-analysis, Graber et al. identified 42 published reports testing interventions to reduce the likelihood of cognitive errors, including educational interventions, training focused on improving clinical reasoning and decision-making skills and analysis, and interventions to provide cognitive "help," including integrated decision-support and informatics tools. All had minimal impact. This does not discredit the idea, however, that cognitive de-biasing may be effective at reducing error if more effective strategies can be developed. Many will argue that merely recognizing that a bias is present may be sufficient to guard against its potential effects. Ongoing work is focused on educational interventions, the use of electronic point-of-care decision-support algorithms, and metacognition techniques (again, "thinking about thinking," including reflection and mindfulness). There may also be a role for group-decision strategies, such as those that take place commonly in the form of interdisciplinary clinical conferences, as well as environmental and cultural interventions, including the simplest possible (but powerful) intervention whereby physicians remind themselves to remain skeptical and to stay open to new information, even after they believe that the correct diagnosis has been well established.

MIND-WANDERING: SPONTANEOUS "INTERNAL" OR "TASK-UNRELATED" THOUGHT

As you were reading this chapter, you may have found that periodically your attention to the text had lapsed, and instead your mind had wandered to other areas of thought—such as making your grocery list, anticipating a future vacation trip, or even reminding yourself about a future task you didn't want to forget. In fact, mind-wandering is a very common occurrence for all people, and studies have shown that this sort of "task-unrelated thought" occupies up to 40% of most people's waking hours!

Psychologists have been studying the phenomenon of mind-wandering and other types of spontaneous thought, such as creative thinking and daydreaming, fairly rigorously in recent years, and there are a few things about it that suggest it might be a factor in the types of radiologist errors that occur most commonly. First, mind-wandering occurs spontaneously, without warning, and is very difficult for people to avoid, even if they try hard to do so and even if they have received special training to improve concentration and reduce mind-wandering. Further, people are generally not aware that their attention has lapsed and do not realize that they have been mind-wandering until sometime after the fact, or unless something prompts their awareness—such as being asked what they were just thinking about. Psychologists call the ability to detect one's own level of awareness "meta-awareness," and it is known to be quite poor where mind-wandering is concerned. Experimentally, a mind-wandering state can be predicted in the presence of activity within certain areas of the brain, which is a topic we will return to later.

Of particular interest regarding the question of mind-wandering and perceptual errors in radiology is the observation from recent neurocognitive experiments that there is a phenomenon of "perceptual decoupling" that is associated with mind-wandering, a mental state wherein people are apparently so focused on their internal train of thought that they become relatively insensitive to all outside stimuli from the senses. During these moments, people are "decoupled," perceptually, from the outside world; they show decreased activation in sensory areas of the brain and make more errors on assigned tasks. This perceptual decoupling can happen even during a task requiring a great deal of focused attention, such as while reading this book, or, if you are a radiologist, perhaps even while you are interpreting a mammogram, chest x-ray, or CT scan. As alluded to earlier, working memory capacity, which can be thought of as the amount of information a person is able to keep in mind during the performance of a task, is also known to be affected by mind-wandering; during episodes of spontaneous, task-unrelated thought, sensory experiences are known to be much less likely to be stored in working memory for conscious use when a person's attention is later refocused on the task. As it happens, testing has shown that different people have different levels of working-memory capacity, and those with more working-memory capacity tend to have less negative impacts from mind-wandering; however, even having an unusually large working memory does not negate the effect completely.

Considering this experimental evidence, it seems unavoidable to question whether it is exactly this type of spontaneous, episodic, and unconscious/non-"meta-aware" mind-wandering, in some form or another, that might actually underlie the bulk of radiologists' randomly occurring perceptual errors in practice. This is an active area of research in our own laboratory; as promised earlier, we will return to this topic in a subsequent chapter on future research directions to improve diagnostic error in radiology.

MORE ON FATIGUE

Most people perform less well on mentally demanding tasks when they are fatigued or stressed, and it does not come as a surprise that radiologists make more errors when their performance is tested at the end of a long workday compared to at the start of their day, as was reported by Krupinski et al. in 2017. In a very elegant study, the researchers presented radiologists at the end of their workday with 64 chest images. The images were presented to them twice, once with and once without the addition of a simulated pulmonary nodule. Not surprisingly, the radiologists' accuracy at nodule detection was reduced by fatigue, but the picture was not a simple one: interestingly, this research found that fatigue reduced all of a radiologist's detection events—leading to misses of true-positive lesions and to reduced false-positive errors as well! Fatigued radiologists seem to be generally less responsive to findings on the images, for good or for ill. These researchers also evaluated their tired radiologists for satisfaction-of-search errors (failing to detect a second lesion after a first one is found) but discovered that fatigue had no real impact on the satisfaction-of-search effect. Finally, the researchers found that, contrary to what one might think, the "older" radiologists (i.e., those who had more years of experience since board certification) were much less susceptible to the effects of fatigue; fatigue tended to have less of a negative impact on their accuracy when tested.

RADIOLOGY AND THE LARGER ISSUE OF DIAGNOSTIC ERROR IN MEDICINE

Diagnostic error in the practice of medicine is known to be a common occurrence and is a major cause of patient harm. The National Academy of Medicine (NAM) 2015 report *Improving Diagnosis in Health Care* provided a conceptual model of the diagnostic process as an iterative one, central to the overall process of healthcare (Fig. 2.3), which is

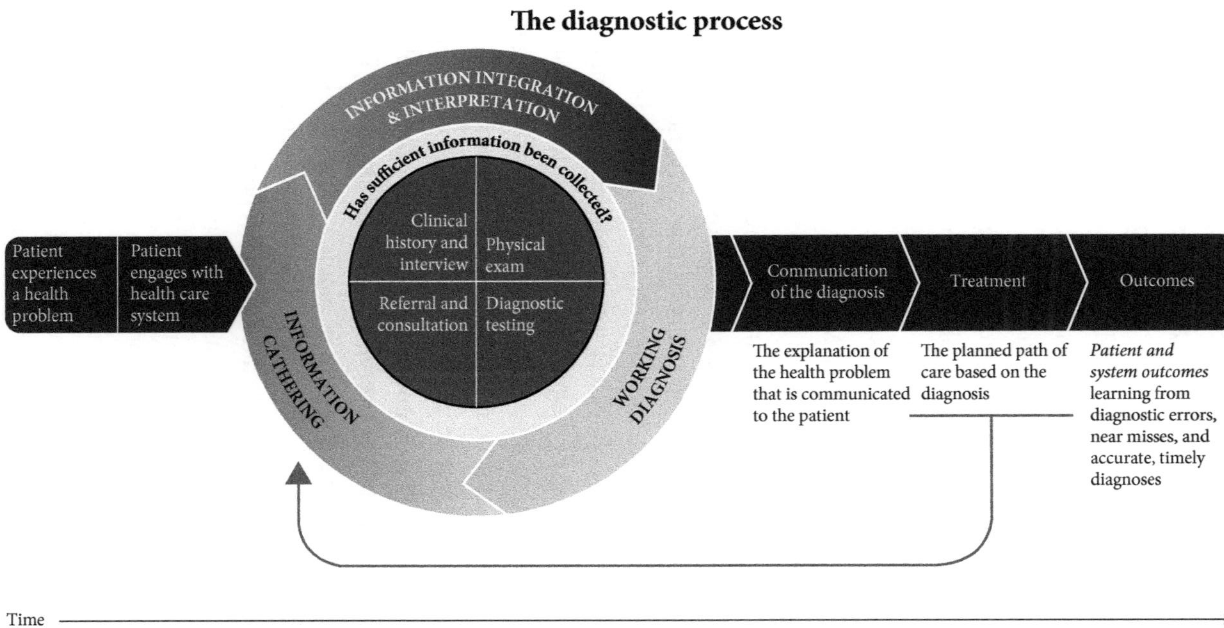

FIGURE 2.3. Conceptual model of the healthcare process, in which the diagnostic process is a central component. Reproduced from the 2015 IOM Report.

imagined as beginning with a patient experiencing a health problem and engaging the healthcare system and ending with a patient outcome, usually involving some sort of treatment. In their report, the NAM called national attention to the problem of diagnostic error, defined the sorts of events thought to constitute diagnostic error, and estimated the prevalence and potential impact of diagnostic error in the United States, which may involve approximately 5% of US adults seeking outpatient care in any given year—that is, every American is likely to experience one diagnostic error in his or her lifetime. The NAM report was the product of years of work involving many dedicated professionals and has led to a broad, national course of action aimed to reduce both the incidence and impacts of diagnostic errors in practice, including within the specialty of diagnostic radiology.

The extent to which incorrect (or at least *discrepant*, as we discussed earlier) radiological interpretations contribute to the overall problem of diagnostic error in medicine remains uncertain. However, given the heavy utilization of imaging in current medical practice, reflecting the current very high degree to which physicians have come to rely on imaging studies to establish a diagnosis, it is likely that the radiological contribution to the overall problem of diagnostic error is substantial. In a recent audit of medical errors resulting in patient harms reported in 2016 to the Pennsylvania Patient Safety Authority under the state's Act 13 reporting mandate, for example, the fraction of cases attributable to radiology was slightly less than 5%. Combined with the Institute of Medicine's estimate of 12 million diagnostic errors occurring annually and the estimate that there are approximately 30,000 practicing radiologists in the United States, this would yield a very rough estimate of 19 or more diagnostic errors attributable to each practicing radiologist, on average, every year.

But there is reason to believe that the actual "error rate" of radiologists in practice is much higher than that, with each radiologist experiencing perhaps as many as three or four errors per day on average, based on multiple studies dating back to Dr. Garland's work in 1949. In fact, no reported measurements of radiologist errors have ever fallen below the approximately 3–4% threshold under any circumstances; in many studies the percentage was substantially greater. Given a crude estimate that the average radiologist workload slightly exceeds 100 case interpretations per day, or 18,000 in a typical year (a figure that is based on older published data and is likely to be underestimated), one can extrapolate a much larger lower boundary for average annual errors for each full-time radiologist each year, which is a number close to 700 (or approximately three or four errors per radiologist per day). I believe the large discrepancy between these two estimates of radiologist errors, with 19 errors per radiologist per year estimated based on extrapolation from reported cases of patient harm versus an empirical estimate of 700 errors per radiologist per year, lends further support to the conclusion that the vast majority of radiologist errors go undetected, or their impacts are mitigated in some way so as to not result in patient harm—or perhaps their effects are obscured by other factors that have greater impact on patient outcomes.

LEARNING FROM PAST FAILURES: KNOWLEDGE GAPS AND PITFALLS

It has been suggested that the triumph of the sciences has been resilience in the face of repeated failure, an idea promoted in the recent and excellent book by Professor Stuart Firestein of Columbia University entitled *Failure: Why Science Is So Successful.* Dr. Firestein points to a pervasive cultural willingness within the scientific community to admit, scrutinize, and carefully analyze past failures and learn from them. While this has not always been the case in the sciences (Max Planck was once quoted as saying that "science progresses one funeral at a time"), and such cultural openness has often been lacking within the medical community, there is some hope that this is changing. Based on Dr. Firestein's reasoning, careful analysis of failed prior attempts to reduce radiologists' errors may well be expected to prove both illustrative and useful.

In the more than six decades since Garland's initial work on radiologists' errors was published, there have been concerted and sophisticated efforts within the radiology community to address the problem. Almost all of this effort has been educational, following the hypothesis that gaps in the knowledge base of individual practitioners are the primary reason for radiologist errors, including apparent failures in perception. A common mantra of radiology educators—*one sees what one knows*—is correctly attributed to the German philosopher Goethe and is commonly recited in connection with continuing medical education (CME). The intensive education of radiologists that begins in their residency training continues more or less unabated after residency with the ongoing updates and retraining (lifelong learning) of practicing radiologists via the various types of CME offerings that exist, including unknown case reviews, journal articles with teaching points and quiz questions, didactic lectures, and in some instances intensive repetition and drill. Whether held in vast hotel ballrooms or university lecture halls, or promulgated in print journal articles or textbooks—and now increasingly via interactive electronic media—a massive ongoing educational effort and industry has developed around the knowledge-gap hypothesis. If one reviews the learning topics and titles from these educational sessions, the most common recurring keyword in CME is "pitfall." Radiologists are constantly admonished to avoid one or another such "pitfalls," or gaps in their knowledge leading to errors, and often are instructed how to do so in great detail by expert teachers. Unfortunately, it has become clear that the educational strategy to eliminate radiologist error, while not entirely without merit, has been pursued quite vigorously for decades across the entire spectrum of diagnostic radiology and has ultimately failed to make any measurable difference. Continuing education to eliminate radiology error would appear to have been a failed strategy—and I think we can conclude that either (1) the underlying knowledge-gap hypothesis is false or (2) the universal aspects of radiologists' training and experience are sufficient such that any incremental gains from small increases in a practitioner's knowledge base have only a very small impact on the error problem, one that is overwhelmed by the larger neurophysiological and neurocognitive factors discussed earlier.

This discussion is certainly not aimed at denigrating or trivializing the value of CME and lifelong learning for physicians of all types; indeed, a good deal of my own professional effort over the past two decades has been spent in providing CME for my fellow physicians within my own narrow areas of expertise. Medical science proceeds apace, and there is always so much more to be learned. But it must be acknowledged that ongoing learning, while in itself a good and valuable thing for the practice of medicine and the specialty of diagnostic radiology, will not in and of itself be sufficient to erase the problem of diagnostic error in radiology. In order for us to make any real progress in this area, future effort and resources must be applied toward better elucidating the *fundamental causes and mechanisms* that underlie the most common types of radiologist error, as well as in developing and carefully testing potential intervention strategies.

ERRORS IN COMMUNICATION

Some of the radiology-related errors that are known to occur are situations where the radiologist has made the relevant observation and arrived at the correct diagnosis, but for one reason or another a failure of communication leads to an ultimate diagnostic error and, often, patient harm. A brilliant quotation on the topic of communication, attributed to the playwright George Bernard Shaw, is "The single biggest problem in communication is the illusion that it has taken place." This idea is certainly applicable in radiology, as lapses in communication, especially between radiologists and the physicians who consult us, are known to be a significant portion of diagnostic error related to the practice of radiology.

When I began my radiology residency in 1988, the primary means of communication of diagnostic impressions between radiologist and other physicians was face to face. The images existed in only one discrete location, on a single piece (or pieces) of 14″ × 17″ acetate film, and teams of physician would come to visit the radiology department each day to look at the images with us, share their knowledge of their patients' condition, and hear our thoughts on what the images revealed. I would argue that there was less misunderstanding back then, due to the give-and-take nature of that communication, but it was a very inefficient process. With the advent of electronic digital imaging, with instantaneous sharing and distribution of images, the inefficiency of a daily (or more often) trip to the basement by every physician has been eliminated, but one cost of this increased efficiency has been a greater risk of miscommunication.

In current practice, radiologists communicate their input on patient care almost exclusively by means of the written word. As a result, radiologists have come under tremendous pressure to provide in their reports all of the actionable information they have to offer and must do so with clarity and succinctness. The report must also be written (and read) under great, and increasing, time pressure. Radiologists are pressured to complete their reports as quickly as possible to facilitate patient care, with little time to explain their thoughts and carefully edit or even proofread for typos. At the same time, the physicians who read the reports are similarly under pressure to do so quickly, often reading them for the first time while in the presence of the patient being cared for, and often within minutes of the report having been finalized and electronically signed by a radiologist (note that electronically signing the report also delivers the final report instantly to the electronic medical record and patient online portals and typically also executes other forms of "delivery," including automatically faxing the report to physicians' offices and clinics).

The written radiology report thus carries the full burden of communication in radiology testing of patients in the vast majority of cases, but that report is a very constrained medium for communication and often leaves much to be desired in terms of its effectiveness at communicating all that there is to know about the results of a given radiological examination, and what the radiologist thinks it might all mean. Radiologists often struggle to decide what to include and what to omit for the sake of brevity and clarity. Verbal discussion, whether in person or over the telephone, between the reader and the writer of these reports, which might serve to clarify any unclear or confusing aspects, only rarely takes place. This systemic state of affairs is essentially a formula for producing numerous communication errors.

THE COMMUNICATION OF UNCERTAINTY

Perhaps the greatest weakness—and the greatest challenge—in radiology report writing in my own estimation is the communication of the radiologist's own level of uncertainty so that the level of confidence intended by the radiologist is understood by the reader. There is plenty of evidence to indicate that we radiologists do not fully appreciate how poorly we are doing in this regard, and also that the clinicians who read our reports do not fully appreciate how incomplete their understanding of the radiologists' intended meanings actually are.

One of the best studies of this problem was published in 2003 by Rhamin Khorasani, David Bates, and colleagues from Brigham and Women's Hospital, who conducted a survey-based investigation to assess the level of agreement that existed among radiologists, and between radiologists and referring clinicians, regarding the level of diagnostic certainty conveyed by particular words and phrases commonly used in radiology reports. While the radiologists believed they were effectively communicating varying levels of confidence versus uncertainty by using phrases such as "possibly," "suggestive of," "consistent with," and even "highly suggestive of," versus phrases such as "questionable," "suspicious for," "worrisome for," or even "unlikely," these nuances were generally lost on their readers. There was concordance on the meaning only of the phrase "diagnostic of." In other words, unless the radiologist was attempting to communicate *complete certainty*, which everyone understood equally, the intended message regarding the radiologist's degree of certainty was not fully received. A more recent study by Andrew Rosencrantz and colleagues at New York University published in 2017 reported similar results. Radiology reports simply do not appear to effectively convey the intended level of uncertainty to the reader.

The situation is analogous to a very funny scene from the popular TV show *Cheers*, which ran from 1982 to 1993, in which the character Frasier Crane reads the opening words of Dickens' *A Tale of Two Cities* to his friends gathered around the eponymous Boston bar. "It was the best of times, it was the worst of times," he says. A long, uncomfortable silence follows. Finally, one of the listeners asks, "Well . . . *which was it*?"

SOME QUICK FIXES

Improving Written Communication

For the past several years, a few of my colleagues and I have been teaching a radiology "report-writing workshop," a learning experience that is offered annually to our own radiologists-in-training (residents and fellows) and that has been shared with other training programs at multiple other academic institutions over the past several years. In this workshop, we assert that the report writer (the radiologist) must keep the needs of the target audience of readers (primarily referring clinicians, but also the patients themselves) foremost. Clarity of thought must be reflected in the report structure and organization, with conscious attention given to such things as the order in which findings are presented, the use of simple declarative sentence structures, and a vocabulary the readers will understand. We emphasize that if the report does not faithfully and effectively communicate our message, we have ultimately failed to serve the patient and have added no value. It is our view that good report writing is an *independent skill* that is best acquired with attention and practice, requiring an investment of both time and focused effort to master. In our own formal assessment, we are encouraged that the workshop appears to have improved the quality and clarity of our residents' reports, at least in the short term, but it is not, in and of itself, a comprehensive solution.

As one might expect, the most difficult of these report-writing skills we attempt to teach is how to faithfully transmit the level of certainty or uncertainty that we intend. As the study by Khoresani et al. showed more than a decade ago, clinicians are often perplexed by our attempts to convey "intermediate" levels of uncertainty in our writing. Since few radiological appearances are pathognomonic for any particular disease, radiologists face this challenge very frequently each day. To accomplish this, we radiologists must find effective ways to communicate our degree of certainty in a forthright, understandable manner if we are going to optimally serve our patients and our clinician customers.

Packing radiology reports with actionable information (signal) and trimming as much extraneous data (noise) as possible, while also making those reports clear, well organized, and completed rapidly enough to satisfy marketplace demands, is challenge enough. But we must also find effective ways to communicate our degree of certainty in a forthright, understandable manner if we are going to optimally serve our patients.

By conveying uncertainty clearly in our reports, we may not be giving our customers exactly what they are asking for—most people would prefer a reliable and simple "yes" or "no" in every case—but instead we would give them what they actually need from the radiological consultation, namely an accurate, understandable, and actionable assessment of the probability that a disease or injury is present, thereby allowing decisions to be made and treatments to be attempted.

In conducting our report-writing workshops over the years, we have also learned that almost any report can be improved by careful proofreading, editing for clarity and succinctness, omitting extraneous verbiage, and maintaining the focus on the clinical question. Similarly, it has often been observed that one may precipitate a misunderstanding by using words, phrases, or abbreviations that are not well understood by the reader. First and foremost, we radiologists *must formulate clearly in our own mind what our level of certainty is* with regard to a perceived finding and/or its significance. Second, we must find ways to explain ourselves within the radiology report in an honest, clear, and straightforward manner, saying what we know and admitting forthrightly what we do not know, and especially what cannot be known, as a result of medical imaging. Uncertainty is an uncomfortable thing for all parties; we like to know that we have found the one "right" answer, but this is not always possible. It can be argued that a great deal of over-testing in medicine happens because doctors are trying to achieve a higher level of certainty than any test will actually allow. And we know that over-testing in medicine can be risky, tends to be costly, and can sometimes lead to more diagnostic error and, worse, patient harm. In contrast, a clear discussion of the inherent uncertainty that remains unavoidable, though an uncomfortable situation, may be just what the doctor ordered.

Use of Checklists and Semistructured Reporting

The use of checklists has been shown to reduce errors of omission in a wide variety of fields, including aviation and critical care medicine, and presumably can also work for radiology. The use of checklists was popularized in Dr. Atul Gawande's 2009 bestseller *The Checklist Manifesto*. The use of structured or "semistructured" reports in radiology, in which the findings are broken down by anatomic categories (see the example in Box 2.1), can serve as a type of checklist to hopefully reduce errors of omission in our reports—for example, by reminding the radiologist to include comments on each organ system. Early published surveys, including our own, have revealed that clinical physicians greatly value this type of report structure (as opposed to the traditional free-flowing narrative structure) for radiology reports, as it helps them to quickly and efficiently extract the information they are looking for and makes the discovery of unexpected findings more apparent to them than traditional prose reports did.

BOX 2.1.

AN EXAMPLE OF A SEMISTRUCTURED RADIOLOGY REPORT

CT Abdomen With Contrast

Clinical History: Abdominal pain

Technique: Enhanced helical CT acquisition was performed, with reconstruction of contiguous 3mm coronal, transverse and sagittal images.

Findings:

Lung Bases: Clear. Heart is not enlarged. No pleural effusion.

Liver, gallbladder and bile ducts: Liver is normal. Gallstones are present, but without inflammatory features, pericholecystic fluid or abnormal gallbladder wall thickening/enhancement. There is no intra- or extra-hepatic biliary dilation.

Spleen: Normal.

Adrenal glands: Normal.

Pancreas: Normal.

Vessels and lymph nodes: No lymphadenopathy; vessels normal.

Kidneys: Kidneys are normal in size and contour and enhance symmetrically. No cysts, masses or calculi, and no hydronephrosis.

Gastrointestinal: Bowel is non-obstructed; no mesenteric masses, no ascites. Appendix is normal.

Bony structures: Multilevel degenerative changes of the spine. No fractures and no lytic or blastic bony lesions are identified.

Impression:

1. Cholelithiasis.
2. No evidence for acute appendicitis.

This type of reporting has become standard in many centers, including my own.

Computer-Aided Detection

An area of intense discussion in radiology today centers around the use of computer-aided detection (CAD) and artificial intelligence (AI) methods for improving radiologists' performance, and possibly even replacing human observers someday, as alluded to in the previous chapter. We will more closely focus on this topic in the final chapter of this book. With advances in computer technology and AI, as many experts point out, it stands to reason that it is only a matter of time before computational algorithms are able to at least augment (if not fully replace) the work of human observers such as radiologists. One notable computer scientist has gone so far as to suggest that hospitals should immediately stop training radiologists, in anticipation of lower demand for their services in the future—although most experts now agree that such advice is probably premature at this point in time. In the brief span of my own career, however, I have seen the ranks of human transcriptionists decimated with the widespread adoption of computerized voice-recognition software (which was very inaccurate and was bitterly derided at the time it was introduced, but rapidly improved), and so the idea that an artificially intelligent machine-learning based computer algorithm might one day, and likely in the span of my own career, be able to "look" at a CT scan and generate a reasonable differential diagnosis for a patient's malady doesn't seem all that far-fetched to me. There are a small number of active startup technology companies (such as Zebra Medical Vision in Israel, see http://www.zebra-med.com/ and Kheiron Medical Technologies in the United Kingdom, see https://www.kheironmed.com/) that were founded with the expressed goal of providing automated, accurate, and timely medical diagnoses using AI, either to augment or to replace human observers, and members of both IBM's Watson and the Google Mind project have also expressed an interest in this problem. In fact, a large database of radiology images was acquired by the IBM group for the purposes of "training" Watson to interpret them.

One can only speculate as the level of their potential future success and how quickly these technologies will advance. Experts at the American College of Radiology and in academia have taken an optimistic view of these ongoing efforts at automation of image-based detection, seeing in them an opportunity for radiologists' performance to be improved, although there is also clearly a risk that radiologists could, at least in part, be disintermediated from the process of image interpretation, with considerable cost savings to the healthcare system, if it can be shown that the accuracy of the machine is equal to or greater than that of the human observer.

Although the future of computer-aided detection and diagnosis may be promising, the past history of CAD in radiology has clearly been a disappointment. For the past several decades, all attempts to augment radiologists with CAD methods, most notably in mammography, have had no actual impact in reducing radiologist errors, and these systems generally perform far less well than human radiologists do. Radiologists have noted a very high degree of false-positive detections from CAD and complained that the CAD devices only detected lesions they had already seen (and fewer of them). But Moore's law predicts a steady rise in the ability of computers to process information, and new "neural network" and "machine-learning" methods are propelling the field forward at a rapid rate. We will return to this topic in Chapter 6.

Technologies based on eye-tracking may someday also aid future radiologists as they do their work, for example by highlighting areas on images that radiologists may have neglected in their visual search, or by providing subtle gaze direction in the radiologists' peripheral vision. This could be done by, for example, having a computer quickly analyze the radiologist's eye movements, identifying areas that were "skipped" on the image, and then subtly altering the brightness of those regions in the image that the radiologist may be neglecting, thereby attracting his or her gaze. Alternatively, this form of guidance could take the form of real-time feedback to the radiologist at the time of interpretation, such as by displaying a colorized heat-map overlay or other annotations to the image, highlighting areas on an image where the radiologist either never looked or did not spend sufficient time on. Prior research has suggested that the radiologist's gaze may linger on abnormalities that he or she is not consciously aware of, so providing such a "visual dwell time map" could be effective at helping radiologists to find abnormalities that they may have actually detected below the level of conscious awareness.

Reducing Interruptions

Almost all radiologists complain about frequent interruptions of their work that occur throughout the workday. Some are essential to their task, such as the need to communicate with referring doctors, but others are more trivial, "nuisance" interruptions that might best be handled by others. Although the actual error-causing impact of interrupting radiologists during their visual search is not known, it stands to reason that reducing all unnecessary interruptions, such as by offloading phone-answering duties, questions unrelated to their interpretive tasks, and so forth to others not actively engaged in image interpretation might be helpful. It is also beneficial for radiologists to have access to the electronic medical record and other needed resources so they can obtain the information they need quickly and efficiently.

Matching Workload to Human Capacity

Much has been written about the hurried, frantic pace of work, long workdays, and other aspects of modern radiological practice, and the available data support the conclusion that fatigued, harried radiologists make more errors of all types. Some of this rapid pace of work and resultant fatigue of radiologists is unavoidable, related to "marketplace" demands for rapid turnaround of radiology reports coupled with high rates of imaging utilization, especially in hospitals and ED settings, coupled with the limited supply of qualified radiologists, especially in some geographic areas. But some of this problem is clearly economic, related to managerial decisions about how much "productivity," usually measured in work RVUs (short for "relative value units," a simplistic metric invented in 1989 that is derived from counting how many radiological examinations are interpreted and the mix of type/complexity present), and thus how much revenue generation is expected of each radiologist on each workday in order to reach a particular revenue target. In most practice settings, a self-comparison of workload, based on RVUs, is made to a common reference standard, such as survey data published by the Association of American Medical Colleges (AAMC) for academic groups or from the Medical Group Management Association (MGMA) for private, non-teaching radiology practices.

Radiology practices generally aim to have their own productivity in line with their peers and will commonly target specific percentile rankings (e.g., a practice that operates at the 50th percentile of the MGMA survey productivity benchmarks would be in the midst of its community-practice peers in terms of how many studies are read by each radiologist each day). These figures are directly tied to the anticipated revenue generated by each radiologist from clinical work—since physician payments under fee-for-service payment plans are generally directly proportional to the total RVUs generated. RVUs per radiologist are thus proportional to radiologists' incomes, and incomes are thus inversely related to the number of individuals among whose efforts that work gets done. There is thus an informal, national consensus among radiology practices as to how many cases a radiologist should be expected to interpret during a typical workday, and this directly translates to levels of financial compensation, including such things as base salary, fringe benefits, and productivity bonuses, which are also tabulated by the AAMC (academic practice) and MGMA (community practice) surveys. Thus, it is generally expected, for example, that a practice where the radiologists interpret study volumes in the 50th percentile of the MGMA's RVU productivity survey should enjoy compensation in the corresponding 50th percentile range of the MGMA's salary survey. Practices generally look to hire additional radiologists when their RVU/radiologist ratios show that they are in the higher workload percentiles of their peers, such as the 75th to 90th percentile, a level where most radiologists throughout the country are reading fewer cases per day, and generally will not consider hiring any additional radiologists if their RVUs per radiologist fall below the 50th percentile in relevant surveys.

It is noteworthy that the absolute number of RVUs, which remain relatively fixed per unit of work, that correspond to each percentile range has been steadily rising over the past several years, such that radiologists must interpret increasingly more studies each day in order to remain in the same relative workload range as their peers, presumably in an attempt to compensate for declining payer reimbursements per unit of work, if their goal is to keep income levels relatively unchanging. The MGMA's 2015 physician compensation survey report revealed only a very slight decline, of approximately 1.3% on average, in radiologist compensation between 2013 and 2014, which was during a period where professional reimbursements to radiologists per unit of work actually declined on the order of 14%, due to a combination of code-bundling, RVU "re-valuing," and other cost-containment measures undertaken by the Centers for Medicare and Medicaid Services. From this disparity we can readily conclude that radiologists are working significantly harder than before and specifically that, on average, radiologists are interpreting more cases per day as a result of perceived economic pressures.

This approach may make good business sense, but there has been very little true scholarship on the workload issue as it relates to radiologist error rates; it would seem worthy of study to clarify from experimental data how radiologists' error rates might co-vary with increasing individual workload, since this would encompass both the speed of practice and cognitive fatigue.

In past years, academic departments have sought to find whether there is an RVU threshold above which academic scholarly or research output is severely compromised, but no one has attempted to find the RVU threshold or range above which errors increase—though it would make sense that such a threshold must exist. Accordingly, the idea of using experimental evidence to guide personnel and management decisions, with the goal of matching the workload of individual radiologists to their human capacity for optimal performance, rather than to desired financial targets, might be a straightforward, if unpopular, partial solution to the problem of radiologist error management.

To take this a step further, it has been widely recognized that the psychological well-being of healthcare providers is a major consideration in the quality of their work. This has led to a widespread focus within (and beyond) the medical profession on the issue of medical provider "burnout," a psychological state associated with both personal distress and greatly reduced overall performance. The AAMC and other national organizations have focused considerable attention on the issue in recent years. To this end, it may also be worthwhile to assess at what individual workload levels radiologists begin to encounter heightened levels of psychological distress (e.g., reports of excess anxiety, burnout, and depressed mood) and to what extent these factors may contribute to radiologists' errors in practice.

SOME CONCLUSIONS

To conclude this section, it is worth reiterating that, when one considers the degree of uncertainty involved in the process of radiological interpretation; the incredible complexity of radiologists' neurocognitive tasks, which predispose to error; the wide variability in process and practice; the pervasive nature of cognitive bias among all physicians (and everyone else); and the challenges of communication that are embedded in our current medical system, it is indeed remarkable that radiologists and the diagnostic process in which they are embedded perform as well as they do! But until such time as the radiologists' task is taken over by robots, the practice of radiology will remain a human endeavor and thus can never be made error-free. Accepting the inevitability of human error and working within that framework is indeed a central aspect of patient safety; finding ways to prevent errors from causing harm is therefore still extremely important.

Many of those harm-prevention strategies hinge on creating a healthy organizational culture, a "culture of safety," including blamelessness, openness, a heightened vigilance and awareness of the risk of error, and other factors. It also makes sense to invest effort into developing checks and balances to reduce the potential harm of errors after the fact, as well as to develop "trigger tools" to facilitate early detection of errors soon after they occur and hopefully before any irreparable harm is done.

Systems-based practice quality improvement strategies designed to better manage radiologists' workload, reduce fatigue and psychological burnout, optimize the speed of radiologists' work (not too fast, but also not too slow), improve overall ergonomics, and reduce interruptions and distractions should all have a positive effect.

More importantly, basic research into understanding the underlying processes of human visual perception as applied to radiologists' tasks are likely to yield new understandings that, in turn, may lead to new, workable, and highly effective strategies to substantially improve radiologists' performance. In the meantime, it would be foolish to pass over or ignore the low-hanging fruit represented by systems-based learning and improvement, including being mindful of the most common forms of cognitive bias; taking steps to reduce the effects of fatigue, interruptions, and distractions; trying to guard against the effects of mind-wandering; using cognitive and memory aids such as checklists; and using whatever means are needed to improve the fidelity of communication between radiologists and the physicians who rely on them.

FURTHER READING

Balogh EP, Miller BT, Ball JR, eds. Board on Health Care Services, Institute of Medicine. *Improving Diagnosis in Health Care.* Washington, DC: National Academies Press, National Academy of Sciences, 2015.

Benson B. Cognitive bias cheat sheet. Better Humans blog, available online: http://betterhumans.coach.me/cognitive-bias-cheat-shet-55a472476b18#.s2j8qandz, accessed January 2017.

Bruno MA, Petscavage-Thomas J, Mohr MJ, Bell SK, Brown SD. An open letter: the radiologist's report in the era of the patient web portal. *J Am Coll Radiol* 2014; 11(9):863–867.

Christoff K, Irving ZC, Fox KCR, et al. Mind-wandering as spontaneous thought: a dynamic framework. *Nature Rev Neurosci* 2016; 17:718–731.

Garland LH. On the scientific evaluation of diagnostic procedures. *Radiology* 1949; 52:309–328.

Graber ML, Kissam S, Payne VL, et al. Cognitive interventions to reduce diagnostic error: a narrative review" *BMJ Quality Safety* 2012; 32(7):535–557.

Kim YW, Mansfield LT. Fool me twice: delayed diagnosis in radiology with emphasis on perpetuated errors. *AJR Am J Roentgenol* 2014; 202(2):465–470. doi:10.2214/AJR.13.11493

Krupinski EA, Berbaum KS, Caldwell RT, et al. Long radiology workdays reduce detection and accommodation accuracy. *J Am Coll Radiol* 2010; 7:698–704.

Krupinski EA, Berbaum KS, Schartz KM, et al. "The Impact of Fatigue on Satisfaction of Search in Chest Radiography. *Acad Radiol* 2017; Sept. 24(9):1058–1063.

ERROR AND ACCOUNTABILITY

A 62-year-old woman faithfully has her annual mammograms on the same day each year. This year, a small mass containing tiny calcifications is detected for the first time, and the diagnosis of breast cancer is confirmed by biopsy. The radiologist realizes that, in retrospect, the cancer was visible on the prior year's images but was not identified. The cancer was still very small at the time of surgical excision, and so this 1-year delay in diagnosis probably did not hurt the patient's chances of survival, although this cannot be known for certain.

In the previous two chapters we focused on the high level of uncertainty inherent to the diagnostic process in general and to the practice of diagnostic radiology in particular, and in doing so we explored the large degree of overlap between error and uncertainty in the practice of radiology. Turning now to the issue of error and accountability in radiology, we must again ask: Under what circumstances is an act or omission of an individual radiologist an "error"? Next, we must consider under what circumstances radiologists' errors are detected (and perhaps try to estimate how often they go undetected) and how radiologists' errors ought to be evaluated and classified. Finally, as a society we must decide on criteria by which *a radiologist ought to be held individually culpable* for errors, whether or not they result in patient harm, and in doing so must consider when a radiologist might thus incur legal liability, professional sanction, remediation—or even more severe consequences, such as loss of employment, revocation of licensure, or loss of clinical privileges to practice within the specialty—as a result of having made an error or a series of errors. It is important to admit that such a determination may not be entirely objective but instead may depend in large measure on the cultural expectations of the milieu in which radiologists practice, and within which errors are evaluated, rather than any inherent quality of the error event itself.

In considering the context-dependent evaluation of errors and what the appropriate response to identified errors should be, it is clear that more than one set of cultural determinants are usually at work simultaneously. For example, many institutions have adopted a "blameless" or "just culture" approach to error discovery and analysis, as is most commonly advocated by quality and safety experts, in which errors tend to be viewed as opportunities for systems and individuals to learn and improve. Others, however, view errors as manifestations of underlying defects in the individuals who are found responsible, thus warranting punishment—which is the more traditional, and individually punitive, approach to errors in medicine.

The prevailing legal (tort) and regulatory regimes, including state licensure boards, hospital administrative boards, and professional societies, are more aligned to this traditional "blameful" way of thinking and are essentially constructed in such a way as to reflect a clear bias toward viewing error incidents primarily as examples of individual failings, perhaps even lapses of individual character. Errors in this model are unanticipated deviations from expected norms or "standards" of professional performance, occurring only as a result of the *negligence of individual practitioners*, negligence for which patients should be compensated and against which the public should be assured a reasonable degree of protection.

In this chapter we will examine these discrepant perspectives in greater detail in the hope of reaching a deeper understanding of error and accountability in diagnostic radiology. We will limit the discussion to focus exclusively on errors that commonly occur in the *routine* practice of diagnostic radiology in the *normal* course of work for *qualified and competent* practitioners, and specifically exclude the special case of impaired physicians (whether the impairment is due to cognitive decline, substance abuse, or mental illness). The impaired physician is an important topic, of course, but is beyond the scope of this chapter.

BLAMELESS CULTURE AND JUST CULTURE, VERSUS THE TRADITIONAL APPROACH TO ACCOUNTABILITY

In the realm of quality and safety improvement in many high-risk industries, including healthcare, the overarching safety culture is designed to be accepting of the inevitability of human failing. It is understood and acknowledged that even potentially preventable errors will inevitably occur.

Thus, the goal is not to somehow achieve perfection in fallible human beings but rather to achieve progressive, relentless, and continuous improvement in the performance of the overall systems in which people work. A central premise to this approach is the idea that patients and society are best protected when all errors are brought to light, analyzed, dissected, and used for both organizational (i.e., systems-level) and individual learning. In this model, most of the errors that radiologists are prone to make, such as simply failing to perceive an abnormality on a scan or radiograph, would be considered to be "blameless"—that is, they would incur no individual culpability. There is therefore no fear of reprisal or shame connected to such an error and no reason for radiologists working in an organizational environment that embraces this type of safety culture to attempt to hide or avoid confronting errors when they occur. To the contrary, this "blameless culture" model has been shown to enhance the self-reporting of errors, which are then reviewed systematically, analyzed rigorously, and discussed openly. Advocates of this approach point out that (1) all radiologists are equally prone to the most common types of error; (2) greater benefit results when everyone has an opportunity to more deeply understand and learn from *all* errors that are detected, and not just from their own errors (whether hidden or revealed), so that future, similar errors are prevented; and (3) this approach has been used successfully in other high-risk/high-reliability industries, notably aviation and nuclear power.

In the prevailing, traditional, and relatively "blameful" model, as we discussed previously, errors have been viewed as primarily occurring as a result of the failures of individuals, which can be avoided if only the operator possesses sufficient expertise and exercises an appropriate level of diligence and effort. Accordingly, persons found to have committed errors are seen to be lacking in knowledge, skill, or character—and thus should be remediated, re-educated, or punished. In this model, errors are cast in a judgmental light and are in essence viewed as shameful, caused primarily by individual lapses, such as laziness, sloppiness, lack of professionalism, or perhaps other more insidious character flaws. In such a punitive organizational culture, when human errors do occur, as they inevitably will, the individuals involved will have a strong motivation to try to hide them; take measures to avoid their discovery; and rationalize, minimize, or deny them. As a result, errors and near-misses by and large are not openly scrutinized, their discovery is blocked or reduced, and little (if any) individual or organizational learning ever occurs as a result. One can predict the results of this vicious cycle: avoidable errors are repeated by multiple individuals with the potential of increasingly severe consequences.

In deciding which approach is the more sensible, it is generally best to be as objective and data-driven as possible rather than relying on "gut" feelings or intuition. Decades of research and a growing understanding of risk in complex organizations have shown that the greatest dangers in a wide range of risky human endeavors are not from individual incompetence, moral lapses, or laziness and that "catching and removing" poor performers within an organization actually does very little to improve the overall system's reliability and safety. While it may be satisfying to assign blame (or find a scapegoat), mete out a "deserved" punishment, and conclude that the error has been appropriately addressed and the problem the error presents is solved, the data show that this strategy is a failing one, and that the satisfaction it provides is false (and short-lived). A core tenet of the quality and safety movement in healthcare has come from simply duplicating the successes of other high-risk industries, such as aviation, manufacturing, and nuclear power. These industries have very clearly turned away from the idea of individual culpability for basic human error and instead focus their efforts on gaining a deeper understanding of errors—identifying as many errors and "near-miss events" as possible through active means, and then studying them and using the knowledge gained to improve the overall system's performance. In these "high-reliability organizations," individual errors are viewed primarily as opportunities to learn and improve, and not as personnel evaluative criteria.

PROFESSIONAL SELF-POLICING VERSUS GOVERNMENTAL AND ORGANIZATIONAL OVERSIGHT

The practice of medicine is certainly a business, and medicine is even (as Dr. Stuart Firestein pointed out in his analysis of failure in the sciences) very much an industry, but it is also a profession ennobled by a fundamental expectation of altruism, its individual practitioners being sworn to put their patients' needs ever before their own and, above all, to do no harm. No other business has such a moral code at its very core. The concept of professional expertise has deep roots in our overall societal culture, dating over multiple centuries when there were well-established "guilds" of experts in many identifiable areas of expertise, including physicians, barber-surgeons, and even astrologers! In this "guild" construct, professional expertise is generally established via a standard mechanism involving a lengthy apprenticeship culminating in some proof of mastery that is acceptable to the supervising/training expert. This general approach still applies in the training of physicians of all specialties, as well as dentists, attorneys, and many skilled trades and is the basic formula for the graduate education curricula leading to doctoral degrees in almost every discipline in the arts, humanities, sciences, and engineering. Once this "apprenticeship" process is satisfied, and only after candidates have proven to their supervisors' satisfaction that they have achieved the desired standard, can they become full members of the professional "guild." The core elements of the guild-apprenticeship model are very much apparent in the processes involved in the training and certification of new radiologists.

In a professional guild, all duly qualified and admitted members are expected to "police" each other, usually by either periodically or continuously conducting an ongoing evaluation of each other's work in some fashion, or at a minimum by reacting to egregious breaches of guild standards. This self-policing is done in order to purge the ranks of substandard members, including those who may be incompetent or unethical. This self-policing is a core function of the professional guild. By virtue of the members carrying out this function, the group as a whole preserves the reputation of the overall profession and also gains marketplace credibility by ensuring that quality standards are universally maintained and the public is protected. The main motivation for the collective self-policing function is, of course, preservation of the guild itself, including the economic self-preservation of each member via protection of the economic value of each member's professional credentials. Failure on the part of the guild members to exercise this obligation faithfully would thus constitute somewhat of a breach of covenant with the public, with risk of losing the public's trust, which carries the risk of losing the economic value of the guild's imprimatur. As this would potentially impact the livelihoods of many people, most of whom have worked for years to establish their credentials, the stakes are quite high. Accordingly, self-policing within all professions is typically seen as an imperative and is often cast as a moral obligation of every qualified expert, from master plumbers to brain surgeons.

In the practice of medicine, accrediting bodies such as the American Boards of Medical Specialties and the Joint Commission (TJC) for hospital and healthcare facility accreditation are direct outgrowths of this self-policing model. In radiology, organizations such as the American Board of Radiology (ABR) and the American College of Radiology (ACR, an organization similar in scope to the Royal College of Radiologists in the United Kingdom) perform these functions, ostensibly on behalf of their membership. The ACR, among other organizations (such as the Intersociety Accreditation Commission), provides recommended practice guidelines and standards and offers practice certification on a voluntary basis, wherein radiology group practices and institutions can achieve a seal of approval by submitting to detailed inspections of their equipment, policies, and practices—including the qualifications of their providers—and, of course, only after satisfactorily remediating any defects that may be discovered.

In more recent decades in the United States, state governments have taken on the role of establishing minimal criteria for entry into a wide range of professions, as well as for maintenance of standing within these skilled fields, by virtue of government licensure conferring the privilege to work.

This practice is thought to protect the public, but it serves the members of the profession as well. State professional licenses are administered by a seemingly exponential growth in the number of state licensure boards. In more recent years the state's reach has extended beyond identifiable and traditional professional groups, such as lawyers, physicians, surgeons, veterinarians, exterminators, and skilled tradesmen, and now (in many states at least) has come to include less highly skilled persons, such as the "technicians" who apply fingernail polish! It is worth mentioning that the growing reach of the state has also had the effect of weakening the self-policing function of the preexisting professional organizations to some degree. However, the self-policing charter of most professional organizations in the practice of medicine in general and in radiology in particular remains intact (at least in theory), and the various accreditation processes themselves typically require some sort of self-policing, often in the form of a peer review system.

Physicians are a highly regulated, closely watched group. They are afforded a great deal of latitude in their day-to-day practice decisions but are generally held accountable when any lapses are identified. In addition to their own professional organizations and state-level licensure boards, as we've discussed, there are also federal agencies, most notably the National Practitioner Data Bank, that collect reports of physician disciplinary actions and malpractice payments (the latter serves as a proxy for the quality of care provided by an individual physician) and make these data available to a wide range of interested parties, including the state licensure boards, individual hospital credentialing committees, and the public. At the "marketplace" level, consumer-facing websites such as "Rate my doc," "Yelp," "WebMD," "Healthgrades,"

"Vitals," "Angie's List," "RateMDs.com," and others have recently been launched, creating an unprecedented level of public scrutiny on physicians as well, with information provided to consumers on the doctor's availability, communication, punctuality, staff friendliness, office environment, and other factors in addition to available information on a doctor's professional standing.

"Some guys from the state board of medicine are here to see you."

Credit: Bob Mankoff, *The New Yorker*. Used with permission.

ESTABLISHING PROFESSIONAL EXPERTISE, OBTAINING CERTIFICATION, AND ONGOING PEER REVIEW

As with essentially all professions and all other medical specialties, there is a lengthy, formalized pathway for persons who wish to become diagnostic radiologists. Typically, a physician-specialist such as a radiologist must first receive a broad, general medical education, achieving certain prespecified standards of achievement in order to receive his or her "first professional degree," generally an MD, DO, or MBBS. After this, there is another, equally intense period of training and supervised practice requiring several additional years of full-time effort and culminating in a series of evaluations and examinations leading to professional certification. This certification—generally provided by the ABR—is a major component to the new radiologist's being deemed fit to practice independently within the specialty, although a small fraction of radiologists practice without the benefit of ABR certification, some repeatedly sitting for and failing the certification examinations in order to make the marketplace claim that they are "board-eligible," a status that is recognized by many hospitals but that is not actually defined by the ABR itself.

Despite this relatively high bar for entry, it is understood that some radiologists, from time to time, will still fall short of the level of acceptable performance sometime after entering practice—whether due to deficiencies in their knowledge or skills, faults in their professional judgment or lapses in professional behaviors, substance abuse, physical or mental decline, or other reasons. This concern is the reason that traditional peer review processes were developed, to serve as a quality control measure. Since quality in the practice of radiology, like other fields of medicine and indeed most highly skilled work, can be difficult or impossible for the layperson to evaluate, it is generally held that the performance and competence of any individual radiologist can be determined only by means of formal review carried out by others of similar background, training, and certification—in other words, peer radiologists who have also had the requisite years of formalized training and uniform evaluation, as well as comparable practice experience. In radiology, this formalized peer review most commonly takes the form of a randomly chosen sample of a radiologist's work, which is retrospectively reviewed by one or more of his or her peers. Those individuals identified as falling short of the minimum performance standard are flagged for remediation or, in some cases, outright removal from practice in order for the specialty as a whole to preserve at least a minimum standard of performance and competency to both ensure public safety and maintain public confidence.

While many experts in the field have advocated that cases to be peer-reviewed should be completely reinterpreted by the second radiologist in "blinded" fashion (that is, without any knowledge of how the case was originally interpreted or by whom), in current practice peer review is done in a nonblinded fashion, where the reviewing radiologist is made aware of the original interpretation and interpreter and merely asked to indicate his or her level of agreement with the original interpretation (the "default" being that he or she agrees with the original interpretation). This type of peer review is organized at the individual hospital department or practice level and is generally facilitated by electronic tools that automatically select a random sample of each radiologist's cases and presents them for review and scoring by his or her peers. Typically these reviews are scored on a 1-to-3 Likert scale, where a score of 1 denotes complete agreement, a score of 2 is deemed a "minor discrepancy," and a score of 3 is considered a "major discrepancy," based largely on the conspicuity of any missed finding or the likelihood that most radiologists (in the reviewer's opinion) would have come to the same or different conclusions about the final diagnosis in the case. The risk of introducing reviewer bias, particularly hindsight bias, into the review process is quite high.

As was mentioned in the prior chapter, the most commonly used peer-review system is known as RadPeer™, created nearly two decades ago by the ACR. In this system, a small sample of a physician's cases are selected for review by his or her peers, and interpretive discrepancies are scored for agreement and also dichotomously for potential clinical impact—those in which the perceived error is deemed likely to have impacted the care of the patient are scored "b," and those where the error, regardless of severity, is thought unlikely to have impacted care are scored "a." Thus a minor discrepancy, which nonetheless may have adversely impacted a patient, would be scored a "2b," for example. A former scoring category of 4, which was used (presumably rarely) to denote unusually egregious errors, was removed by the ACR's RadPeer™ Committee in May 2016. The RadPeer™ scores for each radiologist are tabulated and statistically analyzed and are available to practice managers locally. They are also collected into a massive national database by the ACR so that each radiologist's discrepancy rates can be compared to those of his or her peers within each practice group and to various established national means. It is then up to each practice to decide how to react if and when any radiologist's performance is found to be substandard according to these benchmarks.

It is generally understood that there should be some process of adjudication at the practice level, whether by a formal peer-review committee or by some sort of "appeals" process, so that reviewed radiologists are afforded some due process. This has the potential to cause a great deal of stress and conflict for all of the individuals who are involved in the process: the reviewed radiologists whose errors are detected sometimes feel unduly criticized and threatened, and the reviewing radiologists also experience a wide range of negative emotions. Such reviews are not without potential consequences: for those hospitals that are accredited by TJC, individual practitioner-specific performance data of some sort are typically required as part of ongoing professional practice evaluation (OPPE), which is needed for hospital accreditation, and the RadPeer™ data can conveniently satisfy the OPPE requirements. However, as mentioned previously, the ACR's RadPeer™ Committee has been very clear that RadPeer™ was never intended to be used for OPPE and that even the potential for individual recrimination from peer review is antithetical to the goal of practice improvement through individual and organizational learning from error. TJC also mandates that a focused professional practice evaluation be carried out for any individual physician who is found deficient in OPPE; for this type of evaluation, the individual's performance is more closely scrutinized by practice leaders to determine whether some sort of remediation or practice restriction would be appropriate. In this way, a formal process exists in most centers by which an individual radiologist's interpretive discrepancies, as scored by a small-sample random peer review of his or her cases, can lead to a range of sanctions, including the loss of practice privileges and damage to professional standing; in some cases a radiologist's livelihood may even be seriously threatened.

Knowledge of the range of potential consequences of a peer review case, of course, affects the "upstream" behavior of the peer reviewers. This tends to undermine the validity of the process, one that is already plagued by sampling error and considerable bias. In most cases, concern regarding the consequences of peer review leads to under-reporting of discrepancies out of concern for the well-being of one's peers or one's practice (which, after all, is a small business). In some rare cases, mostly in academia, many of us have observed malicious "targeting" of disliked individuals by peers who are professional rivals or perhaps for other, personal reasons.

Overall, however, and perhaps not unexpectedly, at the national level the aggregate results of RadPeer™ scores in the ACR database reflect a remarkable preponderance of reviews showing complete agreement with the original interpretation—a *stratospherically high level of peer agreement* that is simply not credible. But despite most practices and radiologist groups being able to show their regulators and client hospitals peer-review data attesting to their members' near-perfection, the process of peer review remains a major source of professional dissatisfaction for radiologists, who see it as either a hollow charade, a potential minefield, or both.

There of course is good reason to question the validity of traditional peer review, which has been the "industry standard" for at least 15 years. First, the rate of inter-reader disagreement, as determined experimentally, is much higher than would be predicted by RadPeer™ data. As mentioned earlier in this book, a landmark study by Abujudeh et al., for example, abdominal imaging experts from Massachusetts General Hospital, all of whom were members of the faculty of Harvard Medical School, were asked to interpret a series of studies that, they were told, had previously been interpreted by one of their professional peers, without knowledge of who had done the prior interpretation or what that interpretation had been. Unbeknownst to all was that included in the packet of studies they were reinterpreting in this study were several cases for which they themselves had been the initial interpreter. This "blinded second read" study revealed that these experts disagreed with each other approximately one-third of the time and *they also disagreed with themselves on a surprising 26% of the cases!*

The overarching message derived from extensive prior research in organizational behavior is that when there is even the potential for individual recrimination and punishment for errors, there will inevitably be suppression of error detection. The element of individual feelings of anxiety, shame, and humiliation create a powerful deterrent to bringing errors to light under such a system. It should not be surprising, then, that the national RadPeer™ data reflect only a tiny fraction, less than 1%, of reported discrepancy scores that are above 1 (complete agreement)! Most importantly, there have never been any published data to suggest that the use of RadPeer™ has resulted in any improvement in the performance of a radiology department or practice.

With greater research effort devoted to error detection and analysis in recent years, it has become increasingly clear that *the majority of radiologists' errors are never detected*, as they are not uncovered in the type of random audits that are generally undertaken. This is especially true for those many error events that have no attributable impact on the patient's outcome. There are likely hundreds of thousands of these "lost errors" each year nationwide; their loss from discovery is potentially tragic, as they do not contribute to individual or systems practice improvement or learning at either the organizational or individual level. Thus, undetected errors are likely to recur, and they will not always be without patient harm.

PEER REVIEW VERSUS PEER LEARNING

An alternative to traditional peer review is the "peer learning" approach, as has been advocated most eloquently by Drs. Lane Donnelly of Texas Children's Hospital, David Larson of Stanford, Jonny Kruskal of Harvard Medical School, and others. A peer-learning system is also based on the idea that only a trained and experienced professional can assess the work of another, but unlike traditional peer review, in a peer-learning system the goal of the review process is for teams of professionals to work together to detect *all* errors, study them, and develop strategies for future error prevention and organizational learning, rather than to "police" each other by means of random audits where error rates are determined by data sampling and each error is attributed to a single individual. This approach leverages the advantages of a fully blameless culture and the positive effects of teamwork and human interactions and is also a more comprehensive approach for the discovery of error.

One of the features of high-functioning organizations is that within them individuals generally feel "safe," both psychologically and otherwise, to identify and discuss their own errors. Organizations with this type of culture tend to greatly outperform organizations where the culture is more "blameful," where trust between individuals is lacking, and where group dynamics are thus more dysfunctional. This is why patient safety pioneer Dr. Lucien Leape stated (in his often-quoted 2001 Congressional testimony) that the main impediment to improving the quality of healthcare is the common practice of punishing people who make errors. Leaders in other industries, including the legendary W. Edwards Deming, an early pioneer of manufacturing quality improvement of the 20th century, expressed similar views.

As we've discussed, it is only human nature that individuals working within a punitive system (i.e., one in which errors are viewed as individual failings) can be expected to attempt to hide or minimize errors; the emotions surrounding the traditional peer-review process therefore tend to be negative ones and work against the desired improvement process. In the "peer learning" model, rather than to identify and remediate "outliers," whose error rate from random audit is deemed substandard, the goal is to improve the performance of *all* radiologists within the organization. Methods include an active search for all errors, the rigorous analysis of errors as learning opportunities, and convening educational sessions or meetings specifically to share what has been learned as a result of errors and to collectively learn from them with the goal of improving overall group performance. In such meetings it is also common to highlight successes or "great calls" when they are detected. In such peer-learning conferences, most experts advise that the individuals who made the errors under discussion not be identified—to emphasize that individual blame is not a factor in the peer-learning process. Depending on the nature of the error, peer learning may take the effect of allowing the group to better understand and address the role of cognitive biases in diagnostic error, allow adjustment of the group's collective "action threshold" for a positive finding, or lead to the group's delving into the relevant literature to increase adherence to evidence-based approaches and thus reduce practice variability.

In a peer-learning system, unlike in traditional peer review, random audits of individual radiologist's cases are never used; rather, *all* errors and even "near-miss events" are actively brought forward through multiple means of discovery, *all* are analyzed for cause and effect, and those believed to offer higher learning potential are shared with the full group in greater detail. Those who advocate for a peer-learning program recommend that it not be used alongside a traditional, individually focused "random-audit score-based" peer-review system, as the latter will undermine the benefits of peer learning, and they insist that *no aspect of the information derived from peer learning be used for OPPE or other individual-based practice assessment.* In essence, all members of the team are able to receive rapid and comprehensive feedback for all detected errors so that individual and group learning is fostered in a truly safe, open environment. Such systems are common in other industries and are becoming more the norm for medicine and radiology in other countries, such as the peer-learning program of the Royal College of Radiologists in the United Kingdom. To the extent that prior results are predictive, this sort of "culture shift" in radiology should result in higher overall health-system performance, reduction of diagnostic error, and an enhanced margin of safety and confidence for the public.

THE JUST CULTURE MODEL AS A COMPROMISE

In the just culture model of David Marx (an attorney), individual accountability is suspended for most errors, such as missed findings, which are deemed "human error," but accountability is imposed in cases where individuals take unjustifiable risks or knowingly engage in behavior outside of the accepted procedural norms—such as skipping a mandated "time out" before an invasive procedure is performed or attempting to practice while under the influence of alcohol. In the just culture model, the circumstances of an error dictate the response to that error. Inevitable human failings are thus accepted and *systems interventions* are created to adapt to them. Thus, human tendencies toward forgetfulness or inattention are compensated for by adoption of standardized approaches like checklists and failsafe systems so that patient harm is reliably prevented without the unrealistic demand that somehow human beings should become perfect. Context and process are considered in the evaluation of errors and individuals working in complex systems.

Many of us are persuaded by the pragmatic value of this approach, although it represents a compromise between the traditional and legal/tort culture and the blameless culture of organizational learning, which has been shown to be extremely effective at improving the safety of other life-and-death industries, such as commercial aviation and nuclear power. Indeed, adopting a more purely "blameless culture" model has led to commercial aviation becoming a prime example of a "high-reliability system," with typical safety and performance grossly exceeding anything encountered in medical care (Fig. 3.1). There is therefore good evidence that the highest levels of performance reliability, including maintenance of the highest level of proficiency for skilled individuals, depend on continuously identifying and analyzing errors, developing strategies for error avoidance, and organizational learning, which can only occur in a

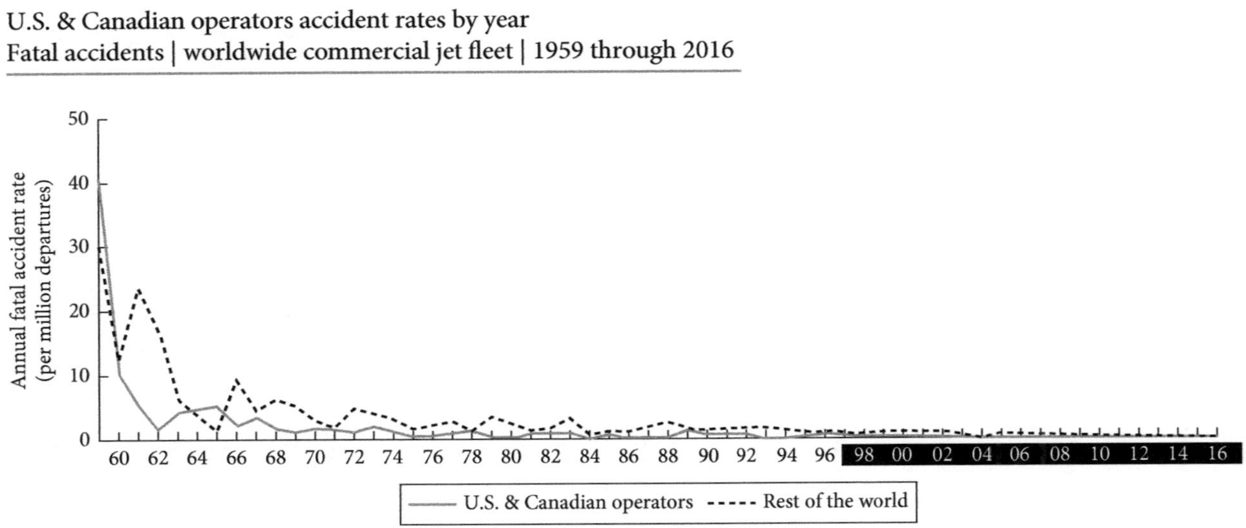

FIGURE 3.1. An example of high-reliability results in airline safety Source: Boeing Corporation Statistical Summary of Commercial Jet Airplane Accidents Worldwide Operations, 1959–2016, available online: http://www.boeing.com/resources/boeingdotcom/company/about_bca/pdf/statsum.pdf

blameless environment where errors can be brought to light without fear of individual recrimination.

Marx's just culture model has a very simple structure: a person found to have made a human error is never punished but rather is consoled, while those individuals who engage in risky behaviors unknowingly, such as by using "work-arounds" to circumvent safety precautions, are counseled and remediated. Only those individuals who engage in overtly reckless behavior (e.g., knowingly taking unjustified risks that are well outside the realm of system/institutional/group norms) are sanctioned or punished. In recent years the just culture model has been gaining traction in radiology, with articles appearing in the major radiology journals and education on its application being offered at the major radiology national meetings. Many experts in the quality and safety realm have embraced the just culture model as an acceptable blend of blamelessness (needed to ensure error discovery) and individual accountability (needed to ensure that the response to bad behaviors is appropriate, for the protection of both the patient and the system).

Despite having been created by an attorney, the just culture model has not been widely adopted in the legal system of the United States or other Western nations. Human errors that would always be considered blameless under the just culture model, such as a radiologist's perceptual miss on an image or occasional lapses of memory or moments of inattention, are sometimes adjudicated by the courts to be examples of *individual negligence*, of carelessness and sanctionable behavior on the part of the individual operator—and as such constitute a cause for compensation to the injured parties, as well as disciplinary action (or at least administrative review) of the involved physician. This is generally when juries are persuaded that the missed finding would not have been missed by a majority of the defendant's peers, or that a procedural lapse in the radiologist's approach may have contributed to the error, thereby constituting a breach of duty or substandard care.

THE LEGAL (TORT) SYSTEM

The legal system is a very specialized application of government oversight into the work of professionals, such as diagnostic radiologists. For the purpose of our discussion, we will concern ourselves primarily with the civil tort legal apparatus, as radiologist errors are generally not considered to be criminal acts under US law.

The legal system is structured as an adversarial system—one side brings accusations and the other side presents its defense—and thus is the very antithesis of a blameless culture. However, it involves well-defined procedures and meticulous and laborious processes and is sophisticated enough to consider the nuances and complexities involved in medical practice. Each side operates with extreme bias in terms of the selection of facts presented in argument, and the goal is to persuade a jury of laypersons, or sometimes a judge, of the merits of each respective position. Accountability for errors is adjudicated somewhat slowly and tediously in the courts, but there is a presumption that every citizen is entitled to such a hearing, although the costs of doing so can be prohibitive. It has been estimated that the cost of medical malpractice litigation exceeds $60 billion annually, thus accounting for approximately 2% to 3% of healthcare costs. Ultimately the money that suffuses the medical malpractice/

tort industry comes from the practice of medicine, distributed via the intermediary of professional liability insurance companies.

In the legal systems of most Western nations, when a radiologist is accused of having made an error that harmed a patient, juries composed of members of the public with little or no medical knowledge are typically presented with expert witness testimony to inform their decisions. These paid expert opinions are often widely divergent, as are the various recollections of the factual events that transpired in the case, as reported by the radiologist in question, ancillary personnel, the patient, and the patient's family members. The jurors are then left to decide whether or not they believe that an error or lapse has actually occurred, and, if so, whether the error was due to individual negligence on the part of one or several of the physicians involved—as is generally asserted by the plaintiff's attorneys—or whether that error was reasonable or justifiable in some way. Accordingly, the attorneys' arguments on each side of a malpractice case often hinge on reviewing the *processes* followed by the radiologist under evaluation: Is there evidence that the error occurred despite an acceptable level of diligence and concern on the part of the practitioner, or is there evidence that the error occurred as a result of deviations from accepted norms, such as rushing through his or her work or failing to attend appropriately to detail? In the era of computerized image displays, information derived from the electronic "audit trail" stored permanently in image management systems may be presented in court to show how much, or how little, time the radiologist spent actually looking at the images before reaching his or her conclusions. The members of the jury may be asked to render judgment on the diligence of the radiologist's process largely based on their own "common sense" of how long that visual-diagnostic process ought to take—despite the scientific evidence that expert radiologists often are able to detect abnormalities on images within half a second (we will explore this issue in greater depth in Chapter 4). The process of litigation is generally plagued by cognitive biases, especially hindsight bias and outcomes bias related to the fate that subsequently befell the patient after the alleged error took place, and it is usually influenced by strong emotions on the part of all concerned.

In the tort system, legal precedent has long established that some errors are "acceptable," in that they do not constitute medical negligence, and it is also generally understood that there is no actual or implied guarantee for patients of a good outcome as a result of their encounter with physicians. But there is also a general legal principle that patients are owed a duty that physicians will apply their full effort, skill, and attention to the cases before them, at least equal to the minimal "standard of care" for the community in which the patient was served. This community standard is, in theory, determined by the local community of the treating physician's peers, but is actually defined as seen through the eyes of the opposing expert witnesses.

The role of the expert witness is worth highlighting: the expert witness is central to the functioning of the entire "medicolegal complex." At the most fundamental level, the truth-finding aspect of the court system is entirely dependent on the performance and "quality" of the available expert witnesses. Again, this is because the legal presumption is that an expert's professional services can be evaluated only by similarly trained and experienced experts in the same field, who themselves practice under similar circumstances.

Yet there are few guidelines and no real training available for radiologists who provide expert testimony; not surprisingly, there is an extremely high degree of variability in expert witness testimony. In practice, any number of expert witnesses can be found to opine that either that the appropriate standard of care was met or was not met for any given incident! Both expert witnesses have sworn to tell only the truth, yet their opinions are generally diametrically opposed. The reasoning of the members of the jury in deciding a case is thus strongly guided by their impression of the credibility of each of the opposing experts. If the expert witness is influenced by a cognitive bias, such as hindsight bias or outcomes bias, he or she may transmit and legitimize that bias in the minds of the jurors. In a practical sense, however, this core aspect of the legal tort system offers a clear avenue to "reform" and improve the legal system's performance. Merely reducing the high degree of variability in expert witness testimony and insisting that evidence-based guidelines be applied in expert testimony would, at least in theory, be sufficient.

Although assignment and adjudication of blame by juries ought to be independent of the nature of the harm done, once culpability is established, the patient's actual health outcome often comes to bear in terms of deciding the extent of the damages owed by the liable party, even though any particular error may or may not have been very closely linked to the patient's final outcome. In rare cases, punitive damages may be awarded if the behavior of a physician is deemed particularly egregious, such as when it appears that attempts were made to hide or alter the evidence of an error in a given case, and it is even possible that criminal penalties may even be imposed if the jury feels that there was some malicious intent on the part of the physician involved.

A BRIEF HISTORY OF MALPRACTICE AND THE PROBLEM OF "DEFENSIVE MEDICINE"

In his research on the history of medical errors and malpractice litigation, Dr. Leonard Berlin found that the legal concept of "medical malpractice" appears to date back to the year 1765 in England, where British legal scholar Sir William Blackstone published a collection of legal commentaries in which he coined the term "*Mala Praxis*," defined as neglect or unskillful management on the part of a physician or surgeon. The first recorded malpractice lawsuit related to a patient's death from a surgical complication. Throughout the 19th century, as Dr. Berlin points out, malpractice cases were almost entirely related to errors in treatment resulting in patient harm, such as a poorly treated fracture or a botched surgery. These were essentially all errors of *commission*, in which the doctor is alleged to have done a poor job in the performance of some active treatment, such as setting a broken bone improperly. Following the discovery of the x-ray in 1895, however, along with the excitement that the x-ray would revolutionize medicine, came a new possibility in the legal system, namely that malpractice lawsuits could potentially be based on errors of *omission*, such as when some abnormality is not appreciated fully on an x-ray image. Such malpractice suits effectively assert a duty on the part of the doctor to perceive abnormalities that are visible in retrospect on images, and they generally allege that a patient has been harmed as a result of a doctor's breech of this duty—such as by having failed to act decisively in the presence of disease as a result of this "miss." Once the specialty of diagnostic radiology arose, it was likely inevitable that radiologists would became the primary targets of this type of lawsuit, and many of these "failure to perceive" lawsuits have been successful.

The legal construct of there being a presumed duty for a doctor to "not miss anything" on radiographic studies expanded throughout the 1950s and 1960s along with advancement in radiological science and seemed to peak in the 1970s, when lawsuits based on errors of omission helped to create a surge of malpractice claims against physicians that has become known as the "malpractice crisis of the 1970s." Increasingly, and especially after the widespread availability of CT scanning, lawsuits were also increasingly being filed against non-radiologist physicians alleging lapses in the standard of care due to the failure of a physician to order radiological studies when such would have been deemed helpful. These lawsuits often alleged that failure to obtain an imaging test for a patient caused an unwarranted delay in diagnosis, leading to loss of the opportunity for early treatment—especially in the diagnosis of cancer, where it is widely believed that earlier treatment will almost certainly lead to a better treatment outcome—an idea that was even more questionable decades ago than it is today. The converse idea, namely that patient harm might possibly also result from over-diagnosis (false-positive testing) and resulting overtreatment (i.e., treatment when a disease is not actually present), did not appear to significantly impact this reasoning. Likewise, the idea that over-testing might actually confuse or delay a patient's diagnosis or even lead to a diagnostic error as a result of spurious test results and an overabundance of irrelevant, misleading, or merely distracting extraneous information is a very new concept in medical thinking and one that has not yet been offered successfully as a defense in a court of law (to my knowledge).

Perhaps surprisingly, it would also seem that defensive medicine behaviors simply don't work to prevent lawsuits. An interesting analysis by Sachs published very recently in the peer-reviewed journal *Annals of Emergency Medicine* asked whether there were measurable physician attributes that might predict (or mitigate) the risk of lawsuits. The researchers evaluated variables such as years in practice, board certification, a pattern of risk-averse practice behaviors (i.e., defensive medicine), relative value units (RVU) production, total number of patients seen, night shift work, patient satisfaction scores, and so forth and found that the only two elements associated with the risk of lawsuits were *increased patient volumes* and *increased years in clinical practice*. In this study, the risk of lawsuits tracked only with exposure, and defensive-medicine behaviors made no difference.

A final point to note about our nation's medicolegal system is that, despite its meticulous, systematic, and rule-driven approach and despite the involvement of many highly skilled experts and professionals in the legal process, most experts agree that our existing tort system does not serve patients harmed by medical errors particularly well, nor has it measurably improved the overall performance of our nation's healthcare system. But despite its overall poor results, fundamental tort reform has proven elusive. There are deep historical roots for current medicolegal practices, and there are also strong financial incentives to maintain the status quo: there is clearly a great deal of money to be made in the "med-mal" industry, and a great many parties stand to gain purely as a result of the grinding of the legal system's various machinations—regardless of whether the plaintiff's claims are ultimately upheld or denied.

From the physician's perspective, it is rather an understatement to say that the experience of interacting with the legal system is a strongly negative one. This, in turn, has fueled a reactive "defensive medicine" crisis in American medicine, typically leading to pervasive over-testing and overtreatment, which has persisted long after the "malpractice crisis" that led to it has resolved. Defensive medicine, simply stated, is when physicians feel pressured to over-test and overtreat patients *primarily* out of fear of litigation—thus leading to wasteful practices driving up costs of care and introducing new risks to patients, as outlined earlier. It is a smaller component of the problem of overutilization than is caused simply by physicians' overall discomfort with the unavoidable levels of uncertainty that persist in medicine despite the availability of hi-tech medical imaging and a plethora of available laboratory tests, as we have discussed earlier; however, the waste-generating and risk-exposure effects of both are additive. Greater availability of electronic methods to provide physicians with evidence-based decision support on a "just in time" basis may help with this problem as well. We will discuss this topic further in Chapter 5.

Despite recent data suggesting that lawsuits are becoming more rare—and that even fewer are ultimately successful for the plaintiff (Fig. 3.2), many doctors still habitually practice defensive medicine, with various deviations from desirable practice behaviors that might be expected to result from a more purely rational or evidence-based approach. These, in turn, clearly carry their own risks of harm to patients as well as the well-documented high levels of wasteful spending in medicine, and it is clear that many of these defensive medicine choices on the part of physicians would qualify as "medical errors" in the context of evidence-based practice.

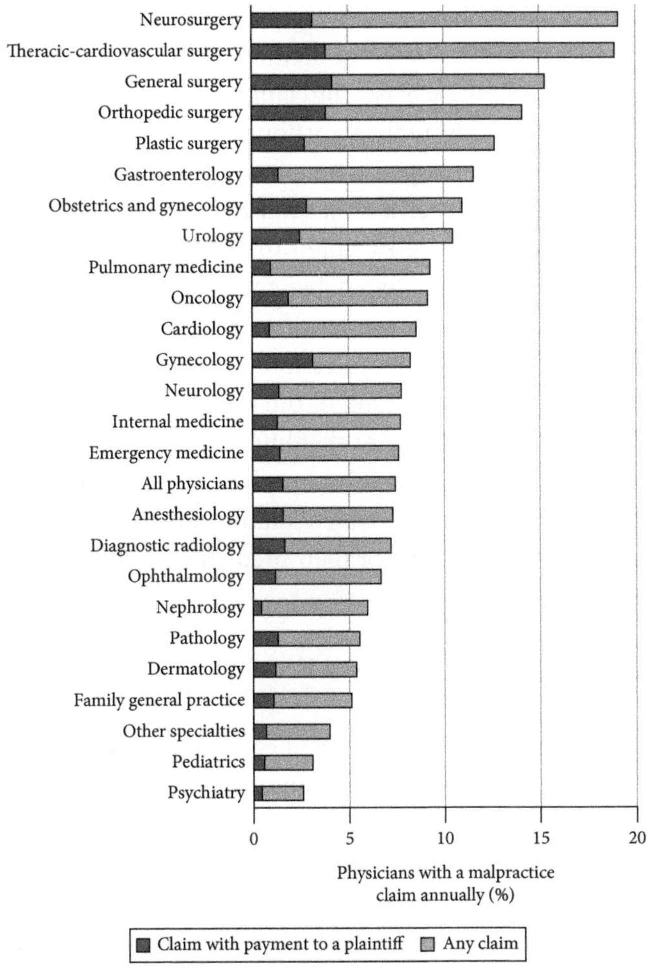

FIGURE 3.2. This graph illustrates that the majority of malpractice claims against all medical specialists, including radiologists, result in no payment. Source: *New England Journal of Medicine*, 2011.

The recent report from the National Academy of Medicine entitled *Improving Diagnosis in Health Care* (2015) compiled a substantial volume of available data on failure in the diagnostic process and concluded that diagnostic errors are so common that *every living person* is likely to experience at least one in his or her lifetime. Many such errors are the result of cognitive biases and faulty reasoning, as we have reviewed previously. Only a portion of diagnostic errors are attributable to radiologist error, since the results of medical imaging are only one of several components of the diagnostic process. The NAM report noted that many of these errors are at least potentially avoidable and preventable. While a small fraction are truly the result of negligent care or reckless behavior by medical professionals, the vast majority are best understood as the result of the sorts of predictable, and perhaps unavoidable, lapses that human beings are prone to make—and patient harm is best understood as the failure of the healthcare systems in which fallible humans work to anticipate and compensate for their predictable risks.

ERROR DISCLOSURE AND APOLOGY

When diagnostic and other medical errors occur, it is increasingly understood that the physician involved has a duty to disclose these errors to patients, and it is customary that an apology is made, along with an explanation of the context and potential consequences of the disclosed error. Many states have enacted laws to exempt such apologies from the content of any subsequent malpractice litigation. However, this has failed to reassure all physicians, and as a result the "medical apology" has not been universally adopted, though few would argue against it on principle. In some institutions, the physician's apology is accompanied by an immediate offer of compensation of some sort, such as waiving a bill for medical services or sometimes even providing a substantial compensatory payment. The apology and voluntary payment applies even for cases that would be considered nonpunishable "human error" under the just culture model, such as a radiologist's failure to perceive an abnormality on an image. In this way, open disclosure, analysis of errors, and systems accountability are all served, with harmed patients receiving some form of compensation, in many cases a great deal more than they might have received via the (very costly and fairly unpredictable) tort system. Published survey studies have shown that patients *expect* to be informed fully and directly when errors occur, and many patients state that they are satisfied once they have been so informed.

Unfortunately, institutional and professional support and resources are often not present for physicians, including radiologists, to effectively accomplish the desired error disclosure, so many radiologists simply do not know how they ought to proceed. In the past, malpractice defense attorneys and medical administrators have actually encouraged doctors to keep silent about errors and harms, to admit to nothing, and thus effectively to "stonewall" patients' inquiries as to what might have gone wrong in their care. This has become known as the "deny and defend" response to error discovery. Such advice seems contrary to the mission of physicians and their duty to their individual patients and represents a "clash of cultures" between the medical culture of compassionate care for individuals based on forming an alliance with the patient and the corporate and legal culture, which, as we noted earlier, is essentially adversarial in nature—thus pitting doctors against the very patients whose welfare they have sworn an oath to place foremost in their work! The approach undermines the doctor–patient relationship at its most basic level, destroying any trust that patients have in their physicians, and thus interferes with the singular task of healthcare, which is to relieve suffering and promote health.

Over 130 professional organizations, including the ACR, endorse the Physician Charter on Professionalism, which explicitly states that patients must be fully informed when medical errors result in injury or lapses of care. Thankfully, it has been increasingly recognized that the old, adversarial "deny and defend" approach to the medical and ethical issue of error disclosure and apology was simply wrong—and yet effective and practical national guidance on the practice of the medical apology is still lacking in radiology. This has thus far resulted in a very high level of variation in practice, whereby patients who have experienced similar errors are treated very differently depending on where they received their care.

The legal system is also slowly coming around to this idea—as noted earlier, the state legislatures of 36 states have already passed "medical apology laws" intended to foster error disclosure by making apology statements inadmissible in malpractice court, unless (in some cases) the actual facts of a case do not correspond to the content of the apology. Naturally, since individual states are free to draft their own particular versions of these laws, there is great variability between states, and some actually only protect expressions of sympathy and not statements of responsibility, making them of limited (if any) value.

A model system is the "Communication and Resolution Program" of the University of Michigan, which has been in place since 2001. Under this regime, all identified medical and diagnostic errors are fully and openly disclosed directly to patients, in person, by the involved physicians. A sincere apology is made (usually by the physician who made the error) along with a clear and empathetic explanation of the circumstances of the error and its ramifications for the patient's future life and well-being. An immediate offer of compensation of some type is typically made along with the apology and explanation. Retrospective review of this program has shown overall lowered liability costs, shortened resolution times for compensation cases, and at least a 40% reduction in malpractice lawsuits.

Many other notable institutions, such as Stanford University Medical Center, have adopted similar programs and since have also reported more favorable patient outcomes and substantial savings from avoided lawsuits than before—including patients generally receiving more financial compensation than they otherwise would have received under the tort system. Involved physicians are supported in their participation and are given no reason to fear for their continued employment or other threat to their livelihood or their work. It is clear that removing the specter of potential personal recrimination for individual physicians is essential for this sort of program to function. These model programs thus provide both "blameless" elements and full accountability to patients for the outcomes of errors, in an honest and open environment that fosters ongoing trust and preserves and enhances, rather than undermines, the therapeutic relationship between doctor and patient.

Casting physicians as compassionate professionals who are concerned with the well-being of their patients but are nonetheless part of the overall human experience (fallible, yet competent and caring) confers tremendous benefits for all involved parties. Error disclosure in these scenarios is tied to systems root cause analyses and other quality and safety efforts to provide organizational as well as individual learning and process improvement, and thus help to "immunize" the healthcare system from making similar errors in the future. The emphasis on quality improvement occurring as a result of error discovery is also reported to be a particularly gratifying aspect to harmed patients. Some patients have stated that the hope that improvements will flow from their experience that will help prevent future errors and thus protect others from harm adds an element of greater meaning and purpose to any suffering they may have endured as a result of physician errors. Without this type of transparency and commitment to process improvement, there is little hope that the practice of medicine can live up to its own standards of patient-centeredness, and even less that medical care can ever become a high-reliability system.

SOME CONCLUSIONS

In summary, while radiologist and other physician errors, as we know, are common, most of the existing mechanisms in place within the healthcare industry, including physician certification and peer-review processes, organizational accreditation, state licensure reviews, professional self-policing, crowd-sourced consumer ratings, and even the legal system, have failed to address these errors in an effective way in order to enhance the overall performance of the medical system and provide greater protection to patients. Moving away from a focus on individual blame and accountability for error and toward a "blameless culture" approach, thereby leading to broader, systemic process improvement in medicine, holds much greater promise for enhancing patient safety than the traditional approaches applied (for decades) in the past, as these new ways of thinking about error have been decisively shown to be effective in transforming other industries, such as commercial aviation, into safe, high-reliability systems.

Adopting peer learning at the individual level and robust error discovery to promote organizational learning at the systems level may provide the greatest impact for future patient safety and error reduction in the practice of radiology. This will require changing the prevailing culture of "blamefulness," in which errors are attributed to the failings of individuals who are held accountable for errors in an atmosphere of shame and recrimination, to a nonpunitive acceptance of the human condition and an expectation that the systems in which humans work must be adapted to known human vulnerabilities. We can expect that the substantial effort required in making this sort of cultural shift will reap the reward of greatly enhanced patient safety.

The just culture model of David Marx may be a useful compromise to help the field transition toward a fully blameless model. The nation's legal system, though flawed and culturally incompatible with the practice of medicine, can nonetheless be aligned to this new approach by means of aligning the content and nature of expert witness testimony to reflect current understanding of medical evidence and the nature of error and uncertainty in radiology. As the tort system evolves, the practice of defensive medicine should give way to increasing reliance on evidence-based approaches, with lower degrees of practice variability (and thus higher quality, reliability, and patient safety). The benefit of evidence-based decision-making in the diagnostic process may be further augmented by wider adoption of effective means of electronic clinical decision support.

Finally, the ethical duty for open disclosure of all errors and lapses directly to patients and their families, regardless of cause and separated from the issue of blame, and sometimes coupled with financial or other compensation for harms done, can only serve to promote the highest standards of quality of care and accountability at all levels.

FURTHER READING

Berlin L. Radiologic malpractice litigation: a view of the past, a gaze at the present, a glimpse of the future. *Am J Roentgenol* 2003; 181(12):1481–1486.

Deming WE. *Out of the Crisis*. Cambridge, MA: MIT Press, 1986. ISBN-13: 978-0262541152

Griffith KS, Marx D. Just culture: a shared commitment. Chapter 7 in Abujudeh HH, Bruno MA, eds., *Quality and Safety in Radiology*. New York: Oxford University Press, 2012. ISBN-13: 978-0199735754

Hanna TN, Lamoureux C, Krupinski EA, et al. Effect of shift, schedule and volume on interpretive accuracy: a retrospective analysis of 2.9 million radiologic examinations. *Radiology* 2018; 287(1):205–212. doi:10.1148/radiol.2017170555

Larson DB, Donnelly LF, Poderesky DJ, et al. Peer feedback, learning and improvement: answering the call of the Institute of Medicine report on diagnostic error. *Radiology* 2017; 283(1):231–241.

Sachs CJ. Malpractice claims: it's a crapshoot—time to stop the self-blame and ask different questions. *Ann Emerg Med* 2018; 71(2):165–167. PMID: 29107402 DOI: 10.1016/j.annemerg med.2017.09.013

SECTION 2

THE SCIENCE OF IMAGING AND PERCEPTION AND THE PROBLEM OF ERROR IN RADIOLOGY

VISUAL AND PERCEPTUAL SCIENCE AND RADIOLOGY

A 42-year-old man falls from a ladder, injuring his left foot, which is noted to be painful and swollen when he sees his family doctor the following day. He has been able to walk on it, but with pain that was not relieved by Advil. The doctor orders an x-ray, which is interpreted by a radiologist as showing soft-tissue swelling over the midfoot, but the radiologist states that no underlying fracture is detected. The man's pain persists, however, and he has worsened difficulty walking, returning to the doctor on crutches after 2 weeks. The family physician sees the patient again and decides to order a repeat x-ray, since the pain has worsened and the swelling has persisted despite conservative treatment. The repeat x-ray, interpreted by a different radiologist, shows the presence of a tarsometatarsal fracture-dislocation, a subtle but serious foot injury that is visible in retrospect but was missed by the first radiologist on the earlier x-rays, although after 2 weeks the fracture has begun to heal and is now more obvious. With this type of fracture, immediate surgery can generally result in a very good outcome; instead, the patient will now face more difficult and extensive surgery and is expected to have a prolonged and painful recovery period and very likely a worsened outcome, with potentially a lifetime of pain and disability as a result of the diagnostic delay.

Radiologists are able to identify an abnormality on a medical image in less than half a second, often before they are even consciously aware that they have done so, and yet are no better at finding Waldo on a "Where's Waldo" puzzle than the average layperson. What neural mechanisms underlie radiologists' astonishing expertise? And what are its limitations? Can the science of visual perception help us to understand radiologists' remarkable abilities and, perhaps more importantly, help us to predict or prevent their failures?

The human visual system has evolved over millions of years and is extremely well adapted for purposes very different from medical image interpretation. Its peculiar (and often amazing) properties enable, but also sometimes hinder, the radiologist's work. A great deal of careful experimentation over the past two decades has helped to elucidate many of the underlying mechanisms of visual search, target detection and recognition, and cognitive analysis, and this research may help to bridge some of the knowledge gaps that have existed in this area. The new research data may also help to explain certain common error types that have been observed to occur in the practice of radiology.

For us humans, most fine visual discrimination occurs in the central (foveal) vision, which is a small fraction of the central (macular) visual field subtended by a very small portion of the retina and which, though small, is served by a

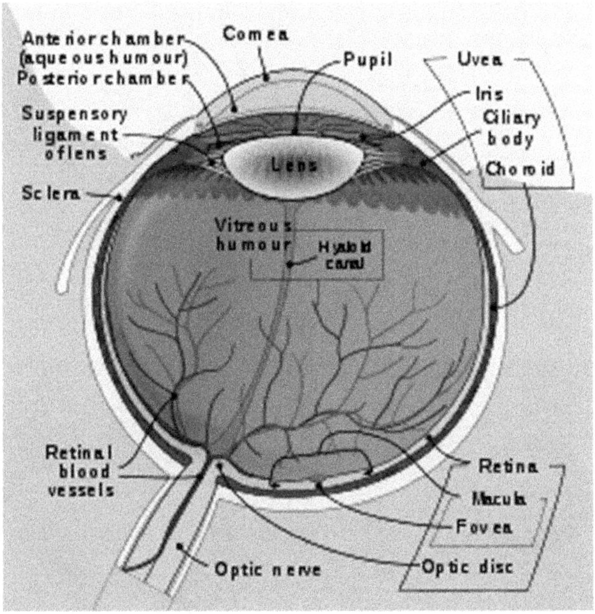

FIGURE 4.1. Diagram of the human eye. Image credit: R.H. Castilhos, public domain, via wikicommons.

very large fraction of the nerve fibers that compose the optic nerve that transmits information to the brain for analysis (Figs. 4.1 and 4.2). The macular region is also allocated much more space in the brain's visual cortex.

The importance of central (foveal) vision was first noted by Leonardo da Vinci, who lived from 1452 to 1519. Da Vinci performed a number of dissections of human cadaver eyes and also conducted experiments using water-filled glass balls. He described the central "line of sight," a ray that connects light from the outside world to the one small part of the inside of the eye for which objects can be seen clearly and distinctly. His remarkable insights into the function of

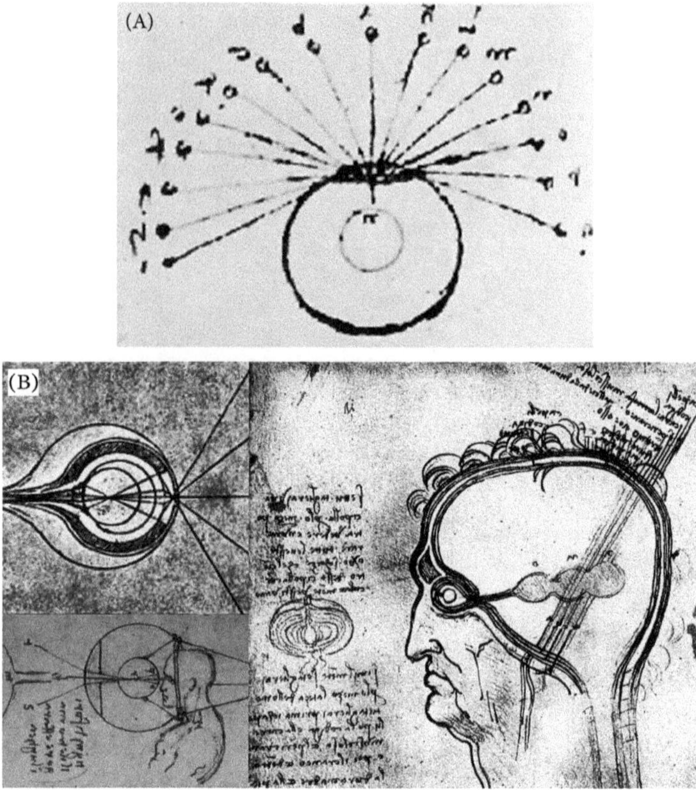

FIGURE 4.2. (A) Da Vinci drawing showing the central "line of sight" and thus making a distinction between high-acuity central vision and peripheral vision. (B) Composite of Da Vinci's drawings regarding the function of the eye in delivering sight to the mind. He showed a particular fascination that the image cast upon the retina was inverted by the eye's lens, an observation later explained by Newton in 1704. Images: Public domain.

the human eye predated Newton's treatise on optics by more than two centuries.

In practical terms, this understanding of human vision suggests that a radiologist's visual task cannot be one of merely passive observation but must instead be an active task, since images must be deliberately "searched," with the radiologist needing to move his or her eyes around an image to concentrate the central visual field on each of many areas of potential interest—one after the other—in order to extract the desired information. Since we know that only a small portion of the central visual field has sufficient acuity for making detailed discriminations and observations, radiologists rarely will be able to have all of the key visual information in their line of sight, so they must hold some of the information from their search in short-term working memory as they complete their search and form an impression. Areas of interest may also be detected in the relatively low-acuity peripheral vision as well, and we now know that there is evidence of an important interplay between foveal and peripheral vision that influences how trained observers scan images.

Technologies have been developed to track a person's eye movements while performing such a visual search, and a great deal of research on radiologists has been based on studying the patterns of their eye movements, including the visual paths taken in scanning the image and the amount of time spent in any particular area of an image. Through years of careful research, it has been learned that the visual search pattern is fairly consistent for each radiologist. We have learned that it can also be guided by their training and personal habits, reinforced by practice, degraded by fatigue, and—importantly—influenced by cognitive biases as well as by the observer's clinical knowledge of the anatomic locations, disease patterns, and types of abnormalities they are expecting to find in their search. Figure 4.3 is an example of a radiologist's gaze-tracking while interpreting a chest radiograph, and Figure 4.4 is a functional magnetic resonance image (fMRI) showing areas of increased regional blood

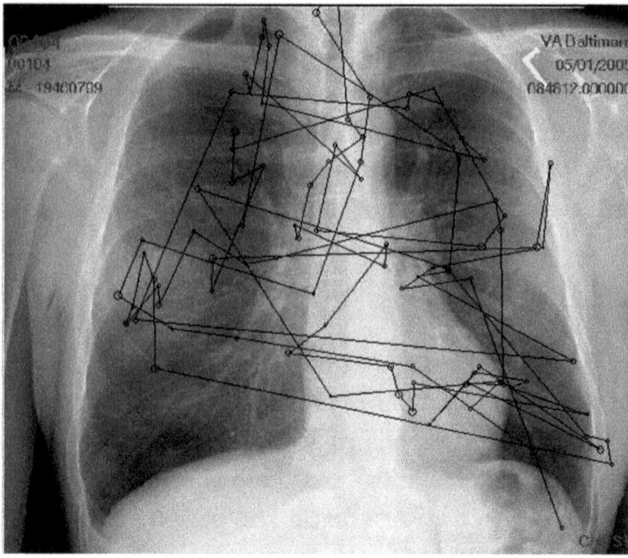

FIGURE 4.3. Gaze-tracking interpreting a chest x-ray. The small dots and circles represent central-vision gaze fixation points of varying durations, and the lines connecting them illustrate the track of the viewer's gaze while examining the image. Note that there are some areas of the image that were never closely examined. Courtesy of Dr. Elizabeth Krupinksi.

flow (which correlates with areas of cerebral activation) in the brain of a radiologist (me) obtained while I was actively interpreting a chest x-ray image.

The fMRI image shows activation in the primary occipital lobe visual-processing areas of the brain, along the visual "pathways," and in the regions of the brain that correspond to the fronto-temporo-parietal network (sometimes referred to as the frontoparietal attention network), including the inferior parietal lobe, dorsal anterior cingulate cortex, and dorsolateral prefrontal cortex, which are specific brain areas that are activated when a person is performing a deliberate task

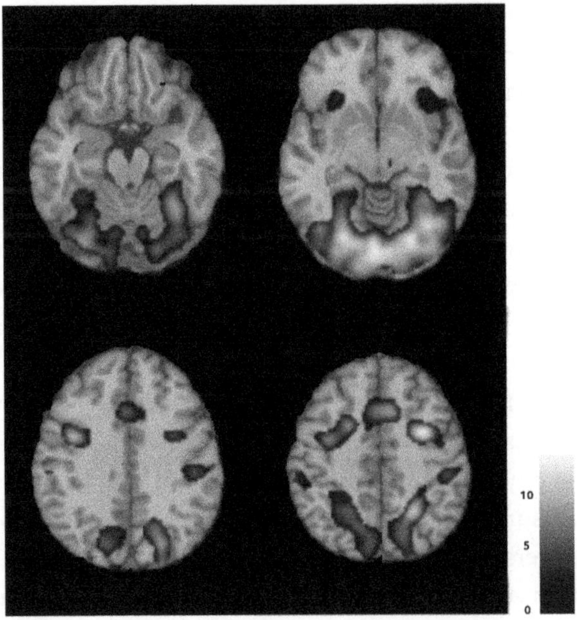

FIGURE 4.4. Functional MR image of the author's own brain, obtained while actively interpreting a chest x-ray image within the MRI scanner, as in Figure 4.3, showing areas of cortical brain activation using BOLD technique at 3.0 Tesla ($p < 0.01$).

requiring focused attention and concentration, and whose disruption in brain-injured patients leads to impaired consciousness. This pair of images illustrate how the approach to medical image interpretation by a trained radiologist is a highly engaged, active mental process in which the images are searched and "interrogated" and their possible significance is systematically considered, rather than a passive process whereby the findings "pop out" visually or are in any way self-explanatory to the observer.

A number of fascinating observations over the past few decades have emerged from these sorts of eye-tracking experiments seeking to understand how radiologists do their work. Studies have shown that radiologist trainees take less time viewing each image as expertise increases. Expert radiologists have fewer gaze fixations and require less time before their gaze becomes fixated on an abnormality than do novices and trainees. There seems to be a natural progression whereby nonmedical observers looking at a medical image will usually have fairly random gaze patterns, not especially influenced by or relevant to the presence of an abnormality, while radiology trainees will show a more organized, slow pattern reflecting their medical knowledge, and expert radiologists will have an almost immediate fixation of their gaze on any abnormalities that are present.

Not surprisingly, an area on an image where radiologists' gaze lingers, even briefly, is highly correlated to the ultimate detection of abnormalities in that location. It may come as a surprise, however, that this observation of "lingering gaze" on areas of images containing abnormalities is also true when the abnormalities are *missed*—in fact, studies have shown that radiologists' gaze may dwell on, and frequently return to, abnormal areas of an image that they later (mistakenly) decide are normal. In one study, it was found that 70% of missed nodules on chest x-rays were looked at for at least 1 second by the radiologist who missed them! This is one of several experimental observations that suggest that there seem to be very important aspects of the radiologists' perception task that are occurring below the level of conscious awareness.

A similar line of evidence for this conclusion was published more recently (in 2016) by Dr. Karla Evans and a group of international co-investigators from the United Kingdom and the United States, who published their results in the journal *Proceedings of the National Academy of Sciences.* These investigators showed that a mere half-second glimpse of a mammogram image was often sufficient for radiologists, but not others, to identify breast cancer cases, even when the radiologist was viewing the mammogram of the *opposite breast* from which the cancer had been diagnosed, and even though the radiologist was not able to say where exactly the abnormality was located on the images. The underlying mechanism whereby radiologists tap into this surprising and perplexing ability is entirely unknown.

In 1975, Harold Kundel and Calvin Nodine published a now-classic paper studying the ability of radiologists to detect abnormalities in situations where performing a visual search was not possible: in their experiment, radiologists interpreted chest radiographs that were only briefly flashed onto a screen. The experimenters projected the chest x-ray images from 35-mm slides using a tachistoscope, a device wherein the brief duration of the image view could be very precisely controlled. Many of the images used were normal chest x-rays, but some included significant abnormal findings, such as lung nodules, pneumonia, a pneumothorax, and even a bullet lodged in the upper mediastinum. Each observer was asked to report what was seen after each "flash" image presentation, with many of the images viewed for as little as 200 msec. Although some mistakes were made relative to the performance of radiologists viewing the same images as long as they wanted, the "flash viewing" detection was remarkably accurate, despite the fact that the abnormalities were not always in the foveal line of sight and no eye movement for image scanning was possible in the timeframe. This result, which revealed at least some limited ability to make diagnoses without any actual visual search of the images, suggests that there may be an initial stage of image evaluation in which experts are able to use information from large regions of an image very quickly, followed by a second stage where a focused search takes place.

This same group of researchers published a study in 1978 in which radiologists searched a set of chest x-ray images, some of which contained a simulated 1-cm tumor, with continuous eye-tracking. The experimenters assumed that, at the typical viewing distance, only a 2.8° radius of the visual field corresponded to foveal vision, and they kept track not only of the gaze path of their subjects but also of the amount of time that a subject fixated on the lesion in order to accurately make the finding. They found that the threshold of time for detection of the tumor was 0.48 seconds. While this is an amazing result in and of itself, perhaps even more interesting was the analysis of the instances where the tumor was *not* detected. About one-third of these errors were considered to be "search errors," where the radiologist simply never looked at the part of the image where the tumor was; one-third were considered to be "recognition errors," where the radiologist found and fixated his or her gaze on the tumor briefly, generally for less than 0.5 second, but subsequently failed to recognize the tumor as an abnormality; and the final one-third were "decision errors," where the radiologist fixated on the lesion for more than half a second but either did not recognize the tumor's features as being concerning or else consciously and actively dismissed the finding as inconsequential. This relative frequency of error types from the "viewing phase" of a radiologist's task has now been replicated by numerous other researchers across multiple imaging modalities. There is probably a valid distinction to be made between "recognition" and "decision" errors, since the former may be reduced by training and experience while the latter may be reduced by both training and efforts at cognitive de-biasing.

Since the late 1980s, the prevailing hypothesis of how a radiologist's image search works is that the task of radiological interpretation begins with the radiologist forming a very rapid global impression, or "gist," which typically requires far less than 1 second to accomplish, often involves peripheral vision, and is in fact the part of the process where most of the initial detection of abnormalities occurs. This rapid "gist" of the image is generally followed by a patterned, deliberate visual search process, wherein the radiologist more closely examines the details of the image using foveal vision in a systematic way, a longer process that can be influenced by a variety of cognitive factors. Eye-tracking studies have revealed that at least two-thirds of missed abnormalities on x-ray images have been shown to have actually attracted close visual scrutiny by the radiologists studied, with missed lesions having been fixated in the radiologists' gaze for nearly as long as those that were correctly detected. This observation suggests that providing radiologists with feedback as to where on an image their eyes had lingered, so that they might take a second look at these areas, might reduce the frequency of missed lesions.

Such a strategy was indeed attempted in an experimental study performed by Dr. Elizabeth Krupinksi and colleagues that was published in 1998. Once a radiologist subject had completed review of a study, the researchers superimposed a colorized "heat map" showing where the subject's gaze had lingered during the search so that the radiologist could re-scrutinize the areas that had drawn his or her attention in the initial review. These researchers showed that such a "guided second look," which had the effect of essentially bringing the radiologist's unconscious notice of possible areas of concern on the images to the level of consciousness, did indeed result in a significant improvement in detection performance without any increase in false positives. Whether this approach could one day gain widespread clinical application for the reduction of perceptual errors in radiology, which as we discussed in Chapter 2 is the most common type of radiologist error, remains to be seen.

Recall that the second most common type of radiologist error, known as *satisfaction of search*, occurs when the radiologist actually discovers an abnormality on an image but then fails to find any additional abnormalities that may also be present. It would appear from several studies, most recently that of Kim and Mansfield, which was published in the *American Journal of Roentgenology* in 2014, that the prevalence of these type of errors is about half that of perceptual errors (i.e., those in which any or all of the abnormalities on the image are simply missed). In one study of detection rates on bone images with multiple lesions, the detection rate of second and third abnormalities on an image was approximately half that of the first. The very name of this error implies a proposed mechanism: the idea that finding an abnormality is in some way psychologically "satisfying" to the observer, and thus the search is prematurely concluded, analogous to the economic decision-making concept of "satisficing," an idea for which Dr. Herbert Simon from the Carnegie-Mellon University won the Nobel Prize for Economics in 1978. Indeed, satisficing, a basic heuristic decision-making strategy in which a search is terminated once some acceptable threshold is met, is likely a general trait of human nature.

Going against the satisficing hypothesis are observations that radiologists' image search time and search patterns do not seem to differ from their usual in the satisfaction-of-search error situation, as well as observations by several researchers that satisfaction-of-search errors do not seem to result from premature cessation of image inspection, but rather seem to involve an unexplained psychological reluctance on the part of radiologists to report additional abnormalities beyond the first, including fewer false positives. A 2017 study, also by Dr. Elizabeth Krupinski and colleagues from Emory University, showed that this satisfaction-of-search effect was mostly unaffected by radiologist fatigue, a factor known to increase the risk of errors of all types. Thus, the cause of this important type of radiologist error remains somewhat of a mystery. One hypothesis under investigation is that seeing the first abnormality somehow captures the viewer's attention so completely as to "crowd out" subsequent abnormalities.

Since image search is a cognitively engaged, active process, the failure of radiologists to perceive or detect an abnormality on a searched image may be influenced by the presence of one or another type of cognitive bias present, as well as by the expectations that radiologists bring to their task. For example, radiologists may be biased by the provided clinical history to search for certain disease entities, to the detriment of other diagnostic possibilities, or they may be biased by their own ideas of what findings are likely or unlikely to be encountered in a given case. We often consider "cognitive errors" as being distinct from those of image perception, but there is clearly overlap as radiologists' search and level of suspicion of findings may well be affected by their preconceived notions. As noted previously, there are more than 40 described cognitive biases that are known to affect both clinical reasoning and information gathering. Some of these are well-known causes of other types of medical error. A few important ones in this regard would include *anchoring bias*, which is when an observer "locks on" to some salient feature that is noticed early and then discounts any conflicting information gathered later; *availability bias*, in which the viewer is influenced by some recent experience that inflates his or her estimation of the likelihood of finding a particular problem; and *confirmation bias*, in which there is a tendency to overvalue evidence supportive of one's hypothesis and discount any refuting evidence. There is also a bias that may be unique to medicine, which has been called *zebra retreat*: a diagnosis that is deemed rare is dismissed out of hand, even though it is actually supported by the available evidence. There is also a well-known *framing bias* or effect, wherein the radiologist's interpretation may be strongly influenced by the clinical question or history provided by the referring clinician—with misleading or incomplete histories leading to findings being overlooked or misinterpreted. It is reasonable to suggest that these biases may act to influence the search process, the incorporation of findings into the radiologist's working memory, or even the radiologist's perception itself, such that some findings may not be remembered, may not influence later reasoning, or may simply not be seen (despite being sufficiently visually conspicuous).

The long years of intense training required to become a radiologist do not change the fundamental toolkit that radiologists use: the human visual system, which evolved over millions of years under evolutionary pressures that did not include radiology and was designed to perform exceedingly well at tasks unrelated to radiological diagnosis. The inherent limitations and hardwired biases built in to this system may have aided the survival of our ancestors in the remote past, for example by helping to decide what stimuli to ignore when too much information is presented, or by helping us make sense of incomplete information by guiding us in making connections and patterns (sometimes where no true connection or pattern exists), or by enabling us to make decisions more quickly under pressure or helping to decide what, if anything, we should learn and remember from any particular experience we may have. But some of these hardwired biases and inherent limitations of the human visual system actually serve to interfere with the application of our human visual "hardware" and cognitive "software" as applied to the novel task of interpretation of diagnostic images. Some of these biases can be unmasked by examining visual illusions. Consider, for example, Figure 4.5, an illusion known as the Fraser Spiral.

FIGURE 4.5. Fraser Spiral. Image: Public domain.

In this illusion, the viewer is strongly convinced that what is being seen here is an inward spiral, when in fact it is not. It is actually a series of concentric circles, which you can verify for yourself by tracing a finger over each of them. What is most fascinating about this illusion, unlike many others, is that knowing the "trick," namely that the image is actually not a spiral at all, has absolutely no effect on the power of the illusion—the viewer simply cannot help seeing it as a spiral, even after surely knowing that it is not. This illusion is thus said to be *cognitively impenetrable.*

A more familiar illusion is the Müller-Lyer illusion (Fig. 4.6). Most people have seen this many times and know that the horizontal line segments are exactly the same length, yet the top one stubbornly appears to be longer than the bottom one. There has never been a completely satisfactory explanation for this.

A similar, less well-known illusion is known as the Shepard's Table, first demonstrated by Roger Shepard

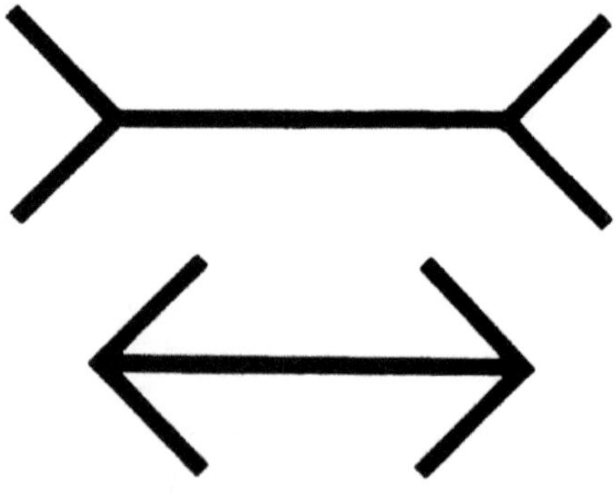

FIGURE 4.6. Müller-Lyer illusion. The central line segments of the upper and lower portions are the same length, although the line appears shorter in the lower portion. Image: Public domain.

(Fig. 4.7). You may need a small ruler to prove it to yourself, but the two tabletops are exactly the same size and shape!

There are many other examples, including illusions of color or grayscale brightness, and even parallel examples in the auditory system in which the brain is similarly fooled, such as Shepard's tone illusion, also named after Roger Shepard, in which a complex sound of overlapping notes of varying intensity is mistakenly perceived as a series of tones continually ascending or descending in pitch, yet ultimately staying the same, and the tritone paradox, an auditory illusion in which pairs of scales may be heard as either descending or ascending by different listeners (when they are

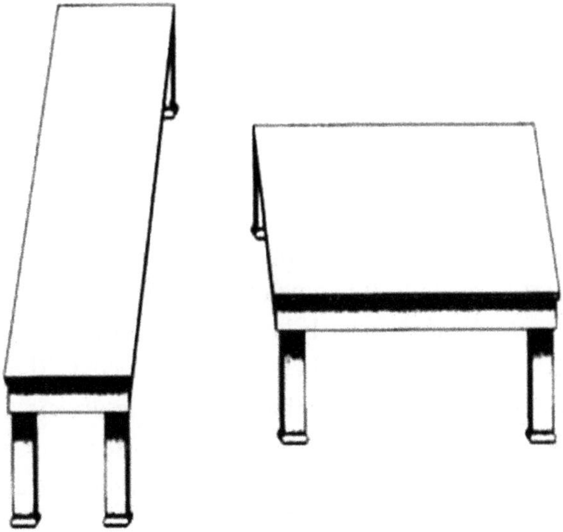

FIGURE 4.7. Shepard's Table illusion. Though they appear to differ visually, both tabletops have the exact same linear dimensions. Image: Public domain.

neither ascending nor descending). These types of sounds, of course, do not occur in nature and are thus unexpected.

One of the best-known radiology examples of counter-intuitive visual phenomena is the "invisible gorilla." In their well-known study, published in the journal *Psychological Science* in 2013, Trafton Drew, Jeremy Wolfe, and colleagues from the Harvard Visual Attention Laboratory showed that most radiologists (83% of those tested) searching a chest computed tomography scan for pulmonary nodules did not see a small image of a gorilla, which was present on the same image as the target nodule, even though the gorilla was larger and more unusual to find than the lung nodule. Eye-tracking showed that at least half of the radiologists looked at the gorilla for at least half a second during their visual search for a lung nodule, and some looked at the gorilla repeatedly, but were later unaware of having seen it. The researchers termed this phenomenon "sustained inattentional blindness," although it almost certainly resulted from the viewers' focused hyperattention to the target nodule at the expense of all other findings present on the images (Fig. 4.8).

Experience has shown that most people (and perhaps everyone) is prone to this type of visual perception error to some degree, in which a fairly large and in retrospect obvious thing is *simply not seen* when it is unexpected. The most common example of this is that automobile drivers all too frequently collide with motorcycles at intersections, apparently due to simply not seeing them while carefully searching their path in the intersection for other cars.

The human tendency to simply not see things that are unexpected, which is so universal as to be considered a fundamental property of human visual perception, may provide the best explanation for radiologists failing to see the

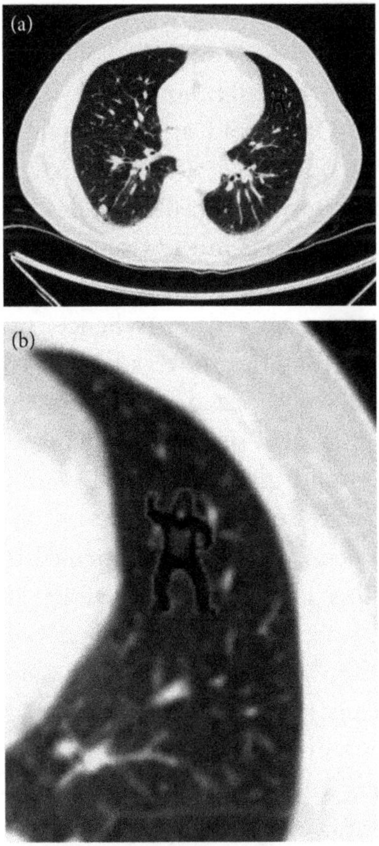

FIGURE 4.8. (Left) A CT lung image showing a small, round, dense nodule in the patient's right lung base, on the lower left side of the image. A much larger image of a black gorilla is present in the anterior portion of the left lung, in the upper right side of the image. (Right) A closeup of the area of the image including the gorilla. From Drew, T., Vo, M.L-H, and Wolfe, J.M. (2013) "The Invisible Gorilla Strikes Again: Sustained Inattentional Blindness in Expert Observers." *Psychological Science*, 24(9):1848–1853.

gorilla. Clearly it is not because the gorilla blends into the background or is camouflaged in some way. The gorilla is there, it is easily visible, it is located right where the observer is looking in fact, but it is *simply not seen*.

Some of the "gorilla effect" is unconscious, of course, but some may be related to individuals' own ability to control and regulate their own attention, vis-à-vis being distracted or preoccupied. One could argue that the presence of a distracting "true positive" lesion interfered with the detection of the "false positive" gorilla. Studies have shown that the appearance of objects in the visual field can be affected by observers' level of attention; it is possible, for example, that the effective conspicuity, or "salience," of a visual stimulus can be altered by the viewer's level of attention. Some researchers have hypothesized that the presence of a finding in the radiologist's field of view can tend to "capture" his or her attention and may be the real reason for so-called satisfaction-of-search errors.

It has also been very well documented that the probability of detecting an abnormality present in a sample of images visually searched is strongly influenced by the "target prevalence" within the sample, which is a measure of how likely it is that any individual image in the "stack" will contain an abnormality. For example, it has been shown that airport screeners will more likely miss the presence of suspicious items in x-ray images of luggage when such items occur only very rarely, which in some cases has prompted the use of "dummy" luggage images containing various kinds of mock-up objects, such as fake knives, guns, sticks of dynamite, or clocks, being added to the mix of actual cases in order to raise the perceived target prevalence and hopefully enhance detection of actual items of interest. Radiologists are equally prone to this "low-prevalence effect" and have similarly been shown to be much less likely to find abnormalities on screening studies where the overwhelming fraction of cases viewed contain no abnormality. For example, in screening mammography it has been estimated that only one to four out of every thousand mammograms viewed will contain a true abnormality, which is a very low target prevalence (the false-positive rate of screening mammography is actually much higher). The mechanism of the low-prevalence effect has not been convincingly established; it may simply be that a low prevalence of abnormalities strongly reinforces observers' cognitive bias that the images they are screening are "almost always normal," such that "no abnormality" is the default interpretation—and is, in fact, correct most of the time. Unfortunately, this bias works to undermine the goal of image-based screening.

In 1985, an eclectic group of researchers from academia, industry, and government who shared an interest in topics related to the perception and recognition of information in medical images formed a new professional organization. It was initially known as the Far West Perception Conference and ultimately became known as the Medical Image Perception Society (MIPS). This is a group broadly dedicated to advancing understanding of the underlying processes involved in medical image–based diagnosis, defining the current state of knowledge in the field, and setting the future research agenda for this broad area (Box 4.1). The group was incorporated in 1998 and now holds a scientific conference/retreat every 2 years to conduct tutorials and workshops for students and to share research results between the members. Since 1994, one of the main priorities of this organization has been to develop psychophysical models for the detection of abnormalities on medical images, including mathematical modeling of the basic tasks of perception, in order to improve the understanding of the mechanisms of perception that are important for interpreting medical images, as well

BOX 4.1.

MIPS GOALS FOR PERCEPTION RESEARCH

Cognitive and psychophysical properties
Detection and discrimination of abnormalities
Perception errors
Search patterns
Modeling human and ideal observers
Computer-based perception and computer-assisted detection
Display and ergonomic factors on image perception and performance
Image processing
Assessment methodologies
Impact of fatigue on observer performance and error

Modified from Krupinski, E. "The Role of Perception in Imaging: Past and Future" in *Seminars in Nuclear Medicine*, vol. 41, pp. 392–400, © 2011.

as to promote methodological research to improve the evaluation of medical imaging systems, such as scanning devices and image displays. MIPS researchers emphasize the ongoing central role of the human observer. This focus seems prudent since, while "machine perception" and the use of artificial intelligence applications may play an increasing role in the practice of diagnostic radiology (the topic of Chapter 6), it still seems likely that the human eye–brain system, in which specially trained and experienced people extract diagnostic information from visually presented images, will remain at the core of the image-based diagnostic process in medicine for the foreseeable future.

An early impact of MIPS led to the design and adoption of the Digital Imaging and Communications in Medicine (DICOM) standard for medical image displays—now universally used throughout radiology—in which the grayscale from which digital images are constructed is "perceptually linearized," which is to say that equal changes in the physical and tissue factors in the patient's body that underlie the image construction are matched to equal changes in the final image luminescence. This is because studies have shown that such a perceptually linearized display actually improves radiologists' ultimate diagnostic accuracy. This is a good example of blending basic human perceptual research with engineering/device design with the goal of improving the final performance of the diagnostic "system," a complex construct that includes both people and machines, so that interpretation accuracy and efficiency are both improved. A number of authors have emphasized this point: the goal of medical imaging research and engineering is not to make "better" images, which may mean different things to different people, nor is it to construct images that are more aesthetically pleasing or even more anatomically correct. Rather, the overarching goal is to improve radiologists' diagnostic performance and improve final patient management. Once the features that are found to be most important to the diagnosis are made as visible as possible, the onus then falls on improving radiologists' ability to perceive those features and correctly understand their meaning.

Most of the research on radiologist and visual-system/display performance has hinged on studies where radiologists are used as observers and the impact on their observer performance is analyzed as a function of various experimental perturbations—ranging from simply altering the brightness or contrast of the display, on one extreme, to more subtle experiments on the other extreme, such as those where small cartoon gorillas are strategically placed on the images in order to determine under which conditions they are, or are not, detected. Increasingly, this type of research is being carried out by large, multidisciplinary teams including radiologists, statisticians, physicists, computer scientists, cognitive psychologists, vision scientists, and neuroscientists, supported by engineers and mathematicians. This underscores both the complexity of the problems being studied and the research questions being asked, their high importance to healthcare and human well-being, and the breadth of expertise needed to even begin to address them.

Perhaps the first of such studies ever reported, that of Birkelo et al. in 1947, announced the then-surprising result that the variability between expert readers was actually far greater than the differences between the imaging techniques and display options being tested! In comparing radiologists to other "expert visual inspectors," such as pathologists and endoscopists, it has become understood that the error rates and accuracy appear similar across disparate visual tasks, suggesting that the limitations in radiologists' performance may actually represent the high-end limitations of the underlying biology that enables human perceptual discrimination.

For example, it is known that the human visual system is attracted more to some features in an image than to others. In any visual image presented to a person, local differences in contrast, color, texture, edge content, symmetry, and orientation of visual elements have been shown to predict the direction of a person's gaze. This is true for images of natural scenes and in a variety of artificially constructed images, including paintings in a museum. The tendency for the eye (and the brain) to be drawn to certain features of an image has been termed "saliency." In essence, it has been hypothesized that the visual system has been primed by evolution to speed visual detection of important features in the visual field by means of their *salience*, such that some features of an image are, in essence, hardwired to appear more interesting, and to draw more visual attention, than others. One of these is a more detailed "texture" of content.

In a fascinating study published in 2014 by Nida Latif and colleagues of Queen's University in Ontario, Canada, and the Royal Ontario Museum of Art, subjects viewed one of several existing copies of an important historical painting, copies that had been created by the same artist, by hand, in the 19th century. These paintings are nearly identical; they use the same colors to depict the same historical event in exactly the same way. However, each is slightly different in terms of regions where there is slightly more granular detail in the image. Subjects wore an eye-tracking apparatus and were given no instructions as to "how" they should look at the paintings. Analysis of their gaze patterns showed that people's gaze was drawn preferentially (although sometimes subtly) to areas of higher spatial frequency. The result was confirmed in a second experiment where a computer was used to manipulate the fine texture of some areas of the image, which again drew the viewer's attention. For the more "salient" regions of the painting—independent of their context or importance to the scene—there was more frequent eye-tracking and longer gaze fixation than other areas.

The evolutionary rationale for the development of this saliency effect is that the information carried on the optic nerve (which has been estimated to perhaps be as much as 10^8 bits per second) is far more than the conscious mind can handle, so some prioritization of attention is simply essential for life. Some of this attentional control therefore happens "automatically," without conscious thought—and the priorities of this "filter" have been established over millions of years by natural selection. For medical images, saliency is closely related but not identical to "conspicuity," those features of brightness and contrast that make an object easier to detect on a complex background. Highly salient features seem to compete for the viewer's attention, and some researchers, notably Howard Egeth and colleagues at Johns Hopkins University, have found that there seems to be a tendency toward a person's visual attention being more or less captured in a "winner-take-all" fashion, followed by a relative inhibition of the return of attention to previously inspected items, back to areas of lesser salience on the image, leading to a cascade of image gaze from the features of highest salience to some (but not all) features of lower salience, with some image features entirely ignored. This mechanism could possibly explain the phenomenon of "change blindness," in which observers are unaware of changes to background features of an image unless they are pointed out to them, at which point they can see them easily. It also might provide a good explanation for "satisfaction of search," errors, as described earlier in this chapter. These well-known phenomena, which affect laypeople as well as trained radiologists, serve to illustrate how hardwired mechanisms built into the human visual system can operate quickly and efficiently—and below the level of the observer's conscious awareness.

A great deal of what radiologists do is under conscious control, however. For example, it is clear that visual search can be improved with practice and can be affected by the expectations or biases of the viewer. A very sophisticated experiment done on 19 healthy volunteers led by Clark and Applebaum at Duke University using electroencephalographic brain mapping aimed to identify which aspects of the visual search process improve with practice. They found that practice can speed up the sensory processing, that there was more rapid direction of the subjects' attention to the target with increased practice, and there was a decreased need for the subjects' recruitment of brain resources to identify or discriminate a positive target from distractors. These results match the experience of radiology residents in training, who over time progressively find the search task to be easier and "more automatic" and find that they are more readily able to recognize abnormalities with greater and increasing certainty, and with greater ease, suggesting a decreased need for cognitive effort as their level of expertise increases.

Comparing the performance of expert radiologists to that of novices clearly shows that experts are able to detect lesions on medical images with a much lower level of attention, and this speed difference is very apparent during "stack viewing" of cross-sectional images. The expert radiologist rapidly scrolls forward and backward through a set of consecutive two-dimensional images across the third dimension, in a more or less constant speed, rather than viewing each two-dimensional image systematically and thoroughly before switching to the next image in the sequence. Some researchers refer to this type of viewing behavior as "drilling." Often, the radiologist's attention is directed toward a particular organ (e.g., liver, lungs, adrenal glands) on sequential images that are rapidly switched, first one direction, then another (e.g., "drilling down through the liver"). The radiologist's attention will then shift to another organ, and the process is repeated, with the same technique applied. In essence, with this stack-viewing technique, the radiologist views the same set of images many times, fairly rapidly, in both forward and reverse sequence and in multiple planes (e.g., cephalad to caudad, anterior to posterior) in each pass with the attention directed to one or another region of the image. It is interesting that, in this context, the image motion itself does not appear to capture the radiologist's attention, as it normally would in other real-world situations.

An interesting observation in this regard is that most people presented with a "stack-view" series of images and instructed to search for a "target" in the stack, such as a single circle among a stack of squares or a single digit among a stack of letters, will more often miss the target when it is presented very early in the stack sequence. Thus, one might predict that radiologists, being hardwired in the same way as everyone else, would similarly be more likely to miss an abnormality that was present only in the first one or two images in such an image stack. This human tendency to miss a target that appears at or near the beginning of an image sequence might then perhaps become an identifiable pitfall for radiologists using the "drilling" method of image interpretation—analogous to missing a finding that is tucked into the corner of an x-ray image. As it happens, however, abnormalities that appear early in a stacked-image sequence, such as a small lung nodule visible only on the first slice of a CT scan of the abdomen and pelvis, are only more likely to be missed by novice trainees and have been shown to actually not be any more likely to be missed than lesions on subsequent images when viewed by expert radiologists. This observation appears to be an example of how the brain's hardwiring can sometimes be overcome by training, experience, and the development of higher levels of expertise.

Another interesting result using fMRI to evaluate brain activation in working radiologists was published by Harley et al. in 2009, in which the brain activation patterns of expert and trainee radiologists at work were observed (as in Fig. 4.4), searching for differences in the brain activation patterns that accrue with experience. It was found that with increasing expertise there was indeed a change in brain activation patterns: the subjects' fusiform cortex became more active and the lateral occipital cortex less active with increasing expertise. This finding suggests that the development of expertise might involve developing new neural representations and pathways or using different cognitive processes to perform the same tasks.

It would make sense, therefore, to conclude that there ought to be detectable changes in the brains of radiologists that correspond to the development of their expertise. To search for clues of this type of neuroplastic change, two neuroradiologist brain researchers at the University Hospital of Basel, Switzerland, Sven Haller and Ernst Radue, designed a study where they used fMRI to analyze the brain activation patterns of 12 radiologists, as compared to 12 laypersons, while viewing either chest radiographs or grayscale electron microscopy images, with the task of determining whether those images had been altered using Photoshop, to create a small blurred area somewhere on the frame. These image modifications were subtle, were fairly randomly distributed, and were a little challenging for people to find.

The electron microscopy images were chosen because they are similar to radiographs in many ways, including in their grayscale and the amount of detail present. Also, neither type of image is generally encountered as part of everyday life, so neither should elicit any particular emotion or memory within the subjects. Only "normal" x-rays were used, and the radiologists were not asked to make any diagnoses on the radiographs, since there would be no way to compare that activity to laypersons. Haller and Radue examined the neuronal activation patterns associated with the visual processing of the images in the subjects' brains, comparing the radiologists to the lay subjects to find out whether the radiologists' expertise and training led to their using different brain areas and mechanisms to perform the visual tasks than did the control subjects. They also compared how long it took for the subjects to find the image alteration, if present (or how long it took them to simply give up and conclude that the image had not been altered). The researchers also measured the subjects' accuracy in that discrimination. The results were published in the journal *Radiology* in a paper entitled "What Is Different About a Radiologist's Brain?" Quite a lot, as it turned out!

First, the radiologists showed greater activation with the radiological images than did the control subjects, especially in the middle and inferior temporal gyri, medial and middle frontal gyri bilaterally, and the left frontal gyrus, brain regions that have been associated in prior research with visual attention and memory retrieval. Activation for the nonradiological images was almost identical for both groups. The radiologists were indeed using somewhat different brain areas to complete the task, showing stronger activation in the left-dominant and more posterior parietal lobe, whereas the nonradiologist subjects showed the strongest activation in the right-dominant and more anterior parietal lobule ($p < 0.001$). These areas have been shown to be involved in creating a mental representation of objects in two and three dimensions and the mental rotation of complex visual stimuli using spatial working memory, which tends to more strongly involve the brain's left hemisphere. It is therefore possible that the experienced radiologists, but not the control subjects, tended to immediately create a "rotatable" mental representation of the radiological images, one that involved greater recruitment of working memory, and that this allowed the radiologists, but not the controls, to be in a position to mentally rotate the presented objects from the images.

The researchers noted that the ability to mentally rotate objects from visual images would be a particularly useful skill for radiologists to develop, since radiological images are often created from patients who are rotated away from standard positions, causing variability in the relationships between anatomic structures. The differences seen in the brain activation during this experiment suggest that the radiologists' visual attention was strongly attracted to the radiological images, which were then subjected to mental rotation and comparison to memorized "reference features" of prior radiological images in the subjects' memory, even though that was not needed (nor was it helpful) for the particular experimental task. Nothing like that was observed for the nonradiological images, which both groups appeared to process similarly. Of course, neither group would be able to draw upon a memory store for the control images, which were equally unfamiliar to both groups. Both groups showed similar (fairly low!) accuracy for the task of finding the Photoshopped image alterations on the images, and both groups reported that they found the task to be difficult. The expert radiologists were only slightly faster in performing the task, averaging a 400-msec speed advantage over controls: the radiologists took about 5 seconds on average, the nonradiologists about 5.4 seconds.

These experimental results reflect the ability of the adult brain to modify itself—at least functionally, if not also structurally—by training and experience, although how this is actually accomplished remains unclear. In the case of acquiring radiological expertise, this study suggests that the brain is able to fundamentally change how it handles a particular set of inputs as a result of learning, even coming to involve different areas of the cerebral cortex than are used by untrained persons or novices. This process almost certainly requires that a person somehow must "repurpose" brain areas and neural information-processing pathways that had evolved to do other tasks to the newly acquired skill and its sharply demarcated (domain-specific) set of tasks.

Similarly specific regional brain activation changes have been observed comparing professional and amateur violinists while performing challenging pieces of music. Structural brain changes related to specific expertise have also been documented in other research. For example, learning-related increases in the volume of gray and white matter of particular regions of the brain have been found as a result of people developing high levels of skill in juggling, and in comparing skilled taxi drivers in London (who need a detailed spatial knowledge of the city) versus "amateur" drivers in London. Similar findings have also been observed in experiments involving animals, such as cats, rats, and monkeys who develop proficiency at novel tasks as a result of repeated training trials. In a fascinating series of studies, Marcel Just and colleagues at Carnegie-Mellon University have recently shown that the neural representations of physics, mechanical, and technical knowledge within subjects' brains rapidly changes with formal instruction and learning, and that it does so in specific, consistent, and predictable ways as the subjects' understanding of the concepts increases.

The observation that the brain plasticity underlying the development of expertise seems to be *so very highly specific* in each example studied provides us with an explanation for the phenomenon we noted at the beginning of this chapter—namely that radiologists are able to find abnormalities on radiographs with almost unbelievable speed, accuracy, and efficiency, but do no better than anyone else at nonradiological visual search tasks—such as finding Waldo on a "Where's Waldo" puzzle.

FURTHER READING

Bernbaum KS, Schartz M, Caldwell RT, et al. Satisfaction of search from detection of pulmonary nodules in CT of the chest. *Acad Radiol* 2013; 20:194–196.

Donald JJ, Barnard SA. Common patterns in 558 diagnostic radiology errors. *J Med Imaging Radiat Oncol* 2012; 56(2):173–178.

Evans KK, Haygood TM, Cooper J, et al. A half-second glimpse often lets radiologists identify breast cancer cases even when viewing the mammogram of the opposite breast. *Proc Nat Acad Sci USA* 2016; 113(37):10292–10297.

Kundel HL, Nodine CF. Interpreting chest radiographs without visual search. *Radiology* 1975; 116(3):527–532.

Mallett S, Phillips P, Fanshawe TR, et al. Tracking eye gaze during interpretation of endoluminal three-dimensional CT colonography: visual perception of experienced and inexperienced readers. *Radiology* 2014; 273(3):783–792.

Nodine CF, Kundel HL. Using eye movements to study visual search and to improve tumor detection. *RadioGraphics* 1987; 7(6):1241–1250.

Samei E, Krupinski E, eds. *The Handbook of Medical Image Perception and Techniques.* Cambridge, UK: Cambridge University Press, 2010. ISBN-13: 978-1107424630.

SECTION 3

STRATEGIES FOR REDUCING RADIOLOGICAL ERROR

5

REDUCING ERROR IN RADIOLOGY

A 70-year-old obese patient with a laparoscopic gastric band for weight loss experiences abdominal pain after consuming a large meal and becomes worried that her gastric band has slipped. She decides to visit the emergency department (ED). The ED doctor orders a computed tomography (CT) scan, which shows slippage of the band and dilation of the proximal pouch of the stomach. However, the doctor goes off shift and signs the patient out to a new emergency physician without reviewing the results of the CT scan. The second doctor also does not check the results of the CT scan but requests a consult from the hospital's bariatric surgeon, who sees the patient and decides to deflate the band in the event that it may be the cause of her pain. At no time do any of the physicians involved look at the CT images or check the radiologist's report, which stated that the band had indeed slipped out of its desired position. The patient is admitted for observation and has a stormy course; a repeat CT scan reveals a catastrophic rupture of the stomach as a consequence of the slipped band, a rare and generally fatal complication of slipped gastric bands. The patient dies shortly after the second CT scan is obtained.

In previous chapters we explored some of the reasons why the practice of radiology can probably never be made error-free. In this chapter we will turn our focus on the topic of how it might nonetheless still be possible to significantly *reduce* the incidence of error in radiology, which would, in turn, be expected to have a tremendous positive impact on the quality and safety of patient care. In considering the way forward for error reduction in radiology, we ought to begin by trying to formulate a working theoretical framework by which to better understand the problem of error and its various causes, and also by reflecting on what has been tried in the past—a history of mostly failed strategies for error reduction—before moving on to take a look at currently promising error-reduction strategies that are currently under investigation. We will conclude this chapter by looking ahead toward possible future approaches to the problem of radiological error.

TOWARD A "THEORY OF ERROR" IN RADIOLOGY

In Chapter 2 we considered ways to classify, categorize, and measure radiological errors, and it seems only logical to consider how each of the different error types we have previously considered may result from widely differing "root" causes. Each error type should be, at least in theory, attributable to a particular cause, or set of causes, and each of the several root

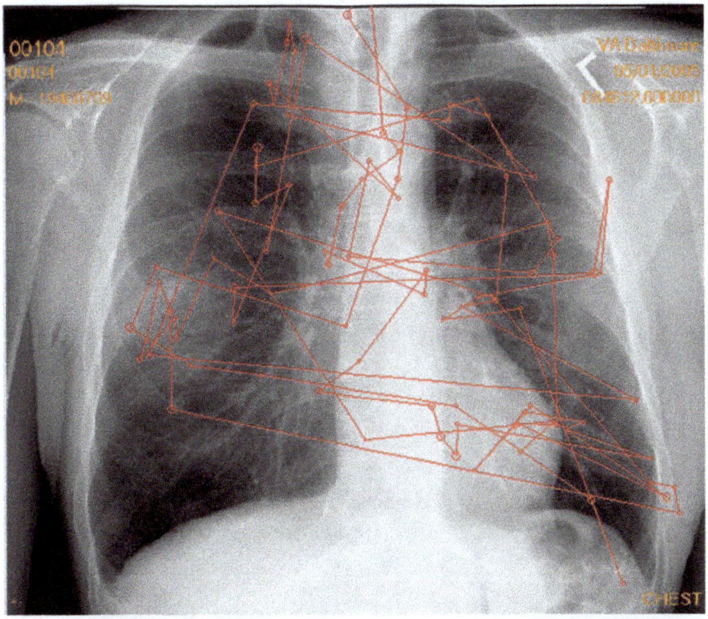

FIGURE 4.3. Gaze-tracking interpreting a chest x-ray. The small dots and circles represent central-vision gaze fixation points of varying durations, and the lines connecting them illustrate the track of the viewer's gaze while examining the image. Note that there are some areas of the image that were never closely examined. Courtesy of Dr. Elizabeth Krupinksi.

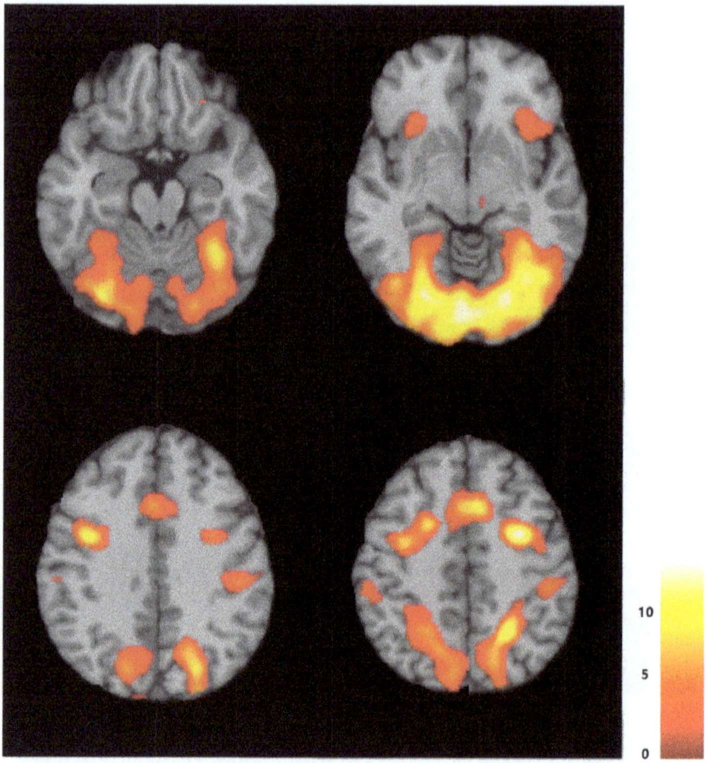

FIGURE 4.4. Functional MR image of the author's own brain, obtained while actively interpreting a chest x-ray image within the MRI scanner, as in Figure 4.3, showing areas of cortical brain activation using BOLD technique at 3.0 Tesla ($p < 0.01$).

causes of error that we can classify should lend itself to a different approach for error prevention and mitigation.

Studies elucidating the prevalence of the various error types and the underlying epidemiology of error, which we have already reviewed, should help to inform our choices as to which types of errors ought to receive the greatest investment of our attention, resources, and effort—since some error types are much more prevalent, and some are considerably more impactful, than others. Yet we also now know that many errors result from *multiple causes working in concert*. These would include systems issues, such as a mismatch of the workload to the staffing level, problematic policies and procedures, technological issues, and the poor design of workflow processes, which, in turn, can often create daunting and myriad inefficiencies (which then result in widespread use of various "work-arounds," created ad hoc). Other sources of error include individual interpreter issues, such as radiologists' fatigue, the fragmentation of radiologists' work, including distractions and interruptions, and still others to the presence of cognitive biases impacting both radiologists' search and recognition processes and their overall "cognitive synthesis" of the cases they interpret. There are also, of course, other highly prevalent "classic" error types that we've already explored—such as errors resulting from satisfaction of search, errors related to the location of the lesion being far from the area of interest or at the margins of the images of a study (sometimes known as "corner" or "edge of film" errors), errors due to gaps in radiologists' knowledge, communication errors, and especially the apparently "pure" perceptual error—those for which the finding is simply not seen. This latter type of error is the most common type by far and when an example is discovered in practice, it often seems inexplicable or simply baffling in retrospect.

To summarize our understanding of error classification and epidemiology into a more basic and comprehensive conceptual framework, we have learned that errors in radiology can broadly be categorized as (1) errors of **how we work** (i.e., errors resulting from myriad systems factors), (2) errors of **how we think** (i.e., errors that result from our cognitive biases and failed heuristics), and (3) errors that result from **how we are made** (i.e., errors that occur as a result of innate properties of our human visual perception and neurobiology). This three-part error model provides, I believe, a much better basic understanding of what we have learned from the science than did the prior prevailing model, which was a more simplistic division of all radiologist errors into only two categories, namely "perceptual error" and "interpretive error." Each of these three broad error categories (systems-based error, cognitive-thinking error, and neurobiologically based error) can be expected to lend itself to a different set of solutions and call for very different types of interventions for error reduction and prevention, as well as for early error detection and mitigation of patient harms when errors do occur. It is our hope—a hope that is, as yet, far from realized—that judicious and broad application of an array of targeted error-reduction strategies, each one aimed at one of these three root causes of the most common types of radiological error, can result in meaningful error reduction and overall performance improvement. To reach this ambitious goal, it would seem as if we will need to first dissect each error type individually and then find a fairly broad range of solutions to apply, each tailored to address one aspect of this broad and complex range of error causes—and then each of these diverse intervention strategies will need to be applied simultaneously.

Errors of How We Work (Systems-Related Errors)

FAULTY PROCESSES AND WORK-AROUNDS

Anyone who has ever been employed by (or even worked as a volunteer in) a large, complex organization can readily grasp the concept of a "faulty process." In healthcare, examples abound: we encounter key work tasks in the healthcare setting that are convoluted, requiring a large number of sequenced steps, including many that do not add value to the work or contribute meaningfully to the outcome. Faulty processes can turn even the simplest task into an inefficient drudgery, and they also can introduce multiple potential failure modes and opportunities for error. A few obvious examples would be the reliance on poorly designed or nonintuitive software to accomplish key steps in a work process, the arrangement of the workplace such that inventories of key supplies are geographically separated from where they are needed or otherwise poorly accessible, and design of processes that include needlessly duplicative tasks and functions, or where there is either poor access to or constrained flow of critical information needed at the points of care. Almost any process that touches the electronic medical record readily comes to mind as an example of such an encumbered or faulty process! In response to occurrences of errors, near-misses, or other safety events that occur, well-meaning managers will often add additional steps to flawed processes that are already overly convoluted, including multiple layers of checks and balances, with the goal of improving safety or reliability. These additional check-steps, however, generally serve only to create added burdens to healthcare professionals and exacerbate the prior inefficiencies of the workflow, thereby creating somewhat of a vicious cycle. Of course, in the healthcare setting, the unique need for extensive human interactions with the sick and injured precludes any of our work processes from *ever* being truly "efficient" in the industrial sense, but faulty process design clearly exacerbates this problem.

Generally, workload demands in the healthcare setting are heavy and time pressure is palpable, so skilled and knowledgeable healthcare workers will typically look toward finding ways to circumvent processes that are inefficient and that slow down their work in an attempt to create greater efficiencies for themselves, ease their task burden, relieve drudgery, and better meet the high-output demands placed on them. In doing so, physicians, nurses, and others will create ad hoc work-arounds of their own (i.e., idiosyncratic alternate work processes that the individuals involved believe will ease and expedite their own workflow). Each of these work-arounds is essentially a unique process in and of itself, and in essence each variation represents an uncontrolled experiment where $n = 1$. Some work-arounds are widely and informally shared and spread throughout the healthcare enterprise, but even when there are such broadly shared ad hoc work-arounds in use, they are typically found to be highly variable in practice (there is no formal training or dissemination strategy for them, after all) and are applied inconsistently across an organization.

The introduction of greater degrees of variability in work processes across the healthcare organization, of course, substantially increases the risk of introducing error as well. Most people who use these sorts of work-arounds each day are not fully aware of their inherent risks, often believing them to have benefits for patients and to increase work efficiency, as studies analyzing workplace work-arounds in healthcare have shown. Needless to say, an organization with a high degree of "*ad hoc*risy" in its daily work tasks and processes is one that is likely to experience a high degree of systems-based error.

One example in radiology would include such things as cumbersome organization of clinical information whereby radiologists might have to check in several different places in the electronic medical record to be fairly sure they know whether intravenous contrast would be safe for a new patient when protocoling a CT study, with work-around shortcuts leading to such patients being placed at greater risk of inappropriate contrast exposure. In many healthcare settings, work-arounds involving postponing and "batching" of tasks can potentially place patients at risk. Sometimes, for example, physicians will take this "batching" approach for tasks which require a separate electronic log-in, in order to save the time and effort consumed by repeatedly logging in and out of the electronic medical record, however such "batching" generally places an unrealistic expectation on individuals to remember what it is they planned to do once logged in, and also causes uneven downstream workflow. An example in the outpatient setting is when an orthopedic surgeon seeing postoperative follow-up patients in clinic chooses to postpone entering their clinic notes and x-ray orders for each patient's next visit until after several patients have been seen, so that fewer time-consuming log-in/log-outs from the electronic record are needed. The surgeon "batches" these work steps to save time and be more efficient, and this work-around has immediate beneftts, as it saves the surgeon precious clinic time and miles of walking up and down the hallways each year between the examining rooms and the workrooms where the EMR is accessed, allowing the clinic to serve more patients every hour. However, it also requires the surgeon to remember (sometimes much later) what transpired in the clinic visit and which x-rays were actually needed for each patient—so this time-saving work-around ultimately depends on human memory, a factor of only moderate reliability! In such a scenario, numerous errors in x-ray orders will inevitably result. Examples abound in other aspects of hospital care as well.

A number of striking examples of such work-arounds, and the problems they can create, have been published in the nursing literature. For example, as a patient safety measure, a nursing process involving the scanning of barcodes for medication administration was found to be faulty: the barcodes were often crinkled and would not scan properly, and the barcode on the package was too small to be read, leading to a variety of ad hoc solutions being applied by nurses to "get around" the barrier. Similarly, when additional safety steps involving "cross-check" barcodes were added to nursing processes, some nurses opted to share their personal barcodes so that when the system called for a cross-check, a second nurse did not actually need to be present. Work-arounds of this type essentially represent a tradeoff of increased operational efficiency and reduced cost in exchange for a higher risk of error.

Addressing the challenge of work-arounds and "*ad hoc*risy" in processes is the fundamental rationale for the Lean systems approach to process improvement, in which any excess steps within any given faulty process are viewed as examples of *waste* and the need for work-arounds is essentially eliminated by fundamental process redesign in order to reduce or eliminate the underlying problems that led to the development and spread of various creative work-arounds in the first place and thus to reduce process variability, and with it the risk of error. It is beyond our scope to discuss the theory and methods of "Lean-Six Sigma" in detail here; excellent tutorials are available for the interested reader.[1] Widespread adoption of Lean and Six Sigma approaches in healthcare organizations can provide powerful tools to help address this type of errors, and in recent years most large hospitals and healthcare systems have either developed their own internal Lean-Six Sigma teams to identify and work on these sorts of problems throughout their enterprise or have resorted to the use of outside consultants to carry out process improvement projects using the well-developed Lean-Six Sigma toolkit.

FAULTY POLICIES AND PROCEDURES, INCLUDING "PERVERSE INCENTIVES"

Organizational leadership decisions, which can be broadly thought of as "policies and procedures," can also be faulty, of course, and are often at the root of systems-based errors. Perhaps the most obvious of these relates to the scheduling and organization of work, such as decisions regarding the number of people required for staffing a clinical section, the division of labor within each of several clinical coverage assignments, the length of each radiologist's typical shift, and so forth. Often these decisions are made in response to financial models and goals, incentives, or other business pressures that are established by, or imposed upon, practice leaders (or that can, in some group practices, be self-imposed by democratic process) without sufficient understanding of, or even a good-faith attempt being made to assess, the safe limits of the involved individuals' actual work capacities and without giving sufficient consideration to the risks created by overwork resulting from cognitive and physical fatigue—nor to the risks caused by requiring a highly accelerated pace of work.

Recent research from a group of imaging scientists and clinicians in the Department of Radiology at Emory University in Atlanta, which was published by Arenson in the journal *Radiology* in April 2018, retrospectively analyzed approximately 4,300 radiologist interpretive errors they had identified out of over 2.9 million radiological examinations performed at Emory over the span of a full year. The researchers determined, perhaps unsurprisingly, that significantly more errors occurred later in a radiologist's shift than earlier in the shift, with errors peaking after 10 hours of continuous work ($p < 0.0001$)! (This is a much higher level of statistical significance than is generally achieved in medical/biological research.) Similarly, when radiologists were called on to interpret higher imaging volumes during their shifts, which is both a measure of overall workload and an indirect measure of the *speed of work* (since the time allotted to complete the work was held more or less constant), this "accelerated pace" was also strongly correlated with increased errors ($p < 0.001$). There was also a trend for a greater risk of errors occurring with an increasing number of consecutive days worked, but that observation did not achieve the same level of statistical significance.

Merely increasing the pace of work with other factors held constant has, in previous research, surprisingly not been shown to substantially increase errors, as was demonstrated in a 2003 study by Edwards and colleagues from the Royal Cornwall Hospital in the United Kingdom, which was published in the journal *Clinical Radiology*. In this study, radiologists interpreted a total of 90 radiographs in sets of

30, working at their normal pace for the first set but forcing themselves to interpret the second set 33% faster (i.e., in two-thirds of the time of the first set) and reading the final set at twice their normal speed (i.e., in half the time). The actual rate of interpretation of each radiograph was not standardized; rather, the radiologists set their own original pace based on their personal comfort and habit, and the researchers then set the speed of the "accelerated" readings relative to each reader's own "comfort" baseline. In this study there was no significant difference in the overall accuracy for each set of interpretations even when the radiologists were reading at twice the speed they were comfortable with, although there were more false positives when the radiologists accelerated their interpretations. The results were found to be highly statistically significant ($p = 0.003$).

Clearly, there are not many people who particularly enjoy working at an accelerated pace for hours and days on end, even if they can do so without loss of accuracy. To the contrary, extended shifts with a sustained rapid pace of work can often be very distressing. Long work hours and the need to maintain a frantic work pace, as well as heavy work volumes, which have the effect of essentially crowding out time to pursue one's personal life and self-maintenance, have all been cited as major contributors to radiologist "burnout" and are major reported causes for radiologist career dissatisfaction and job turnover. While these sorts of faulty work policies and leadership decisions in various industries can sometimes result from shortages of skilled workers, in radiology they seem to be primarily driven by the single-minded and steadfast pursuit of organizationally-determined financial targets and incentives. It is noteworthy that other high-stakes enterprises, such as commercial aviation, place strict restrictions on the duty hours of pilots, air-traffic controllers, and other key workers, with the industry absorbing significant additional operating costs by doing so. The same is true of commercial truck and bus drivers. In academic medicine, duty-hours restrictions have been broadly imposed for trainees who provide direct care to patients, such as interns and residents; however, no such restrictions are applied to their supervising faculty physicians, whose work directly generates the practice revenue.

Broadly speaking, financial incentives that indirectly lead to unsafe or unsustainable practices can be thought of as being "perverse" in that they do not actually foster the overarching and long-term aims of the healthcare organization—aims that are most commonly expressed as a mission to sustainably provide safe and high-quality healthcare to patients in a cost-effective manner. Indeed, the costs related to radiologist burnout and job turnover have not themselves been systematically studied as yet, but these may well exceed the cost savings achieved by the current prevailing strategy of pushing the limits of individual productivity as far as is humanly possible. It is my own personal view that the pursuit of specific financially-derived productivity targets, such as relative value unit (RVU) production thresholds per radiologist, at the expense of increasing the risk of errors and of inducing higher levels of burnout and a greater degree of radiologist job dissatisfaction and turnover, is a fundamentally wrongheaded management approach—one that will ultimately undermine, rather than support, the overall (stated) mission of the healthcare organization. But changing this cost-based approach to managing a radiology practice will almost certainly result in lower levels of radiologist compensation, which can, in turn, lead to short-term challenges in the recruitment and retention of radiologists.

There is an emerging literature suggesting that nearly all management-created financial incentives and "pay-for-performance" plans for physicians, though put into place for well-intentioned reasons, can paradoxically serve to undermine intrinsic worker motivation and thus have unintended negative "downstream" effects. Examples of this problem are as abundant as there are formalized financial incentives! Despite this growing understanding of their drawbacks, "cafeteria-style" financial incentive plans of varying complexity have become a component of nearly all radiology group practices, and especially in the academic setting, with various payments incentivizing a wide range of activities to varying degrees, modeled after incentive compensation plans in other industries. These are widely thought to be effective and beneficial, despite the growing evidence to the contrary. The most obvious example, as alluded to earlier, is that radiologists are sometimes financially rewarded for achieving certain productivity milestones—which can have the effect of tacitly encouraging them to work longer hours or at a faster pace than they themselves would generally deem comfortable (or safe) in order to trigger the incentive payment.

Other well-intentioned incentives can backfire in less straightforward ways. Consider the "blood donor's dilemma," a classic example of the paradox of unintended effects of financial incentives. In real-life trials, the number of individuals who are motivated to donate blood has been shown to *decrease* when a monetary incentive is offered. This appears to be because most people choose to donate blood for altruistic reasons, based on their own intrinsic (and noble) motivations. If a monetary compensation is offered, these individuals may then come to feel that the offer of money *devalues* their more selfless motivation; in this way, an intrinsic (powerful) motivator, one that brings the blood donor lasting satisfaction, is thus substituted for an extrinsic (weak) motivator, namely money, which brings them only fleeting satisfaction.

In an example from my own academic radiology department that occurred not very long ago, our abdominal imaging group noted that our trainees were being required to spend too much of their time each day performing mundane tasks, such as answering phones and protocoling CT and magnetic resonance imaging (MRI) studies, and as a result were not spending enough time actually interpreting imaging studies under faculty supervision, which is the core task of their training and education. To help with this situation, a modest faculty financial incentive was created in an attempt to motivate faculty radiologists to protocol more studies themselves, thus freeing up more of the trainees' time for supervised interpretation. In order for faculty members to earn this small financial reward, they would need to "hit a number"—that is, reach a threshold level of abdominal CT and MRI studies protocoled during the incentive "audit" period. It seemed very straightforward, fair, and equitable and was designed to address a departmental priority (teaching) that everyone agreed with.

But, analogous to the blood donor's dilemma, the faculty protocoling incentive had the paradoxical effect of actually *reducing* the number of studies that were protocoled by the faculty. Why? The incentive was designed to increase the average number of protocols completed by faculty members, in order to decrease the average number completed by residents. But as it turned out, while most faculty members were indeed below the threshold number of protocols at baseline, and a few were very far below the threshold, there was a small minority of faculty members who were very far above the desired mean. This small number of faculty members was responsible—at baseline—for the bulk of the protocols completed by the faculty. Once the incentive was put into place, almost all of the faculty members who were below the threshold level of protocol output at baseline faithfully increased their level of protocoling to meet the incentive threshold, while the smaller number who had already been substantially above the target threshold at baseline essentially did just the opposite: they met the new target by *reducing* their protocoling productivity. In essence, the threshold for the incentive payment established a "uniform target" for each faculty member to achieve, which had the effect of "resetting" the expectations of these few high producers, and thereby undermined the intrinsic motivation they had previously to protocol a larger number of studies than the norm. With a "normative" level of protocoling established, these previously "high-output" individuals reduced their protocoling productivity so that it was no longer a high multiple of the target level, as it had been before.

In the design of the incentive, no one had considered the "Pareto Principle" that was in effect for CT and MRI protocoling in the division: prior to the incentive being instituted, *most* of the faculty protocoling work in the division had been done by this small minority of "high-output" faculty members. This particular aspect of our work had been distributed in a highly inequitable fashion, yet no one had been distressed about this inequality prior to the incentive. With the incentive established, however, the protocoling work was now shared much more equally and "fairly" across the entire group, but at a much lower level for these previously high-producing individuals. The result was a net loss in the protocol output of the faculty as a whole, since the lost output from the high-producing minority was not completely compensated by the increased protocol productivity of the majority.

In the end, the small incentive payout was indeed made to almost all faculty members in the division as a reward for reaching their target, but the residents' protocol burden, the reduction of which had been the overarching goal, had not appreciably changed. The incentive was ultimately withdrawn, but the damage was done; now it seems to me that our trainees seem to be devoting even more time to writing protocols than ever before!

All of these are examples of the sort of problems that can result from attempting to manage a complex enterprise, such as a healthcare organization, by means of establishing a set of numerical goals—which in essence amounts to attempting to alter the workings of a system of extraordinary complexity by applying the simplest and least customized of interventions. The quality management guru of the 20th century, W. Edwards Deming, who is considered to be the "Father of Performance Improvement" and creator of the concept of total quality management, and who is sometimes credited with single-handedly turning around the post-WWII Japanese economy, is frequently misquoted as having said "you cannot manage what you cannot measure." In actuality, what he wrote on this issue (on p. 121) in his classic textbook, *Out of the Crisis*, was essentially the opposite: "The most important figures that one needs for management are unknown or unknowable . . . but successful management must nevertheless take account of them." Deming actually described management by quantitative measures alone as being a "deadly disease of management."

While it may seem anathema to administrators, managers, and business school theorists, the evidence seems to be showing us that in order to achieve the goal of improving individual and overall system productivity and performance, we should probably be considering the *outright elimination* of the target-driven (or pay-for-performance) incentive schemes we have become accustomed to using in healthcare, as Deming advocated in 1982, and instead replacing them with something that better resembles Deming's ideal of "intelligent leadership"—that is, a management style that is better informed by an understanding of the basic nature of human beings in the workplace and what motivates them.

In place of traditional monetary incentives, perhaps greater benefit might be realized by using insights derived from the burgeoning field of behavioral economics. For example, managers might instead seek to find ways to leverage workers' *intrinsic* motivation. This might perhaps be accomplished by better aligning short-term operational metrics with the most broadly shared overarching organizational goals, or by working to enhance the perception among the staff that individual work tasks—even those that are mundane or repetitive—nonetheless contribute to achieving the (shared) desired outcomes for the organization and for the patients served in key and meaningful ways, or by emphasizing how they may enhance the safety and well-being of patients. Physicians, as a group, are people who have exhibited high levels of intrinsic motivation throughout their lives and who tend to view their work as a noble calling rather than as merely a means to earn a daily wage. Appealing to their sense of professionalism, meaning, and higher, noble purpose, as in the blood donor example above, may lead to a greater degree of self-motivation to exceed baseline expectations than would an appeal to do so in order to derive some incremental financial gain. Additional advantage may also be gained for the organization by leveraging the interests and passions of individual physicians, for example by attempting to better match work assignments to individual interests and expertise rather than by assuming that everyone in a particular working group is entirely interchangeable. The impact of applying such an unusual set of management strategies on error reduction is only speculative; this approach to personnel management in healthcare in general, and in radiology in particular, is not only unstudied but essentially untried.

FAULTY WORK CULTURES

When an organization's work culture is lacking in trust and teamwork and there is little effective communication, or where individuals are fearful or do not feel safe in their work or in their positions, the work culture is an unhealthy one. Dysfunctional organizational cultures predictably have a substantially increased risk of systems-based errors of all kinds. For example, in work cultures that lack basic trust, there is a greater reliance on hierarchical authority and formalized rules, which are factors that have been shown to lead to widespread systems failures, as well as to a lack of adaptability and resiliency when there are inevitable perturbations from expected routines. Steep hierarchies with large gradients of status and power effectively prevent the two-way flow of information vital to error detection and correction—in other words, information must flow not only from the leaders directing the front-line workers but also from the workers to the leaders. Micromanagement by leaders that is centered around short-term results and a pervasive sense of pessimism or cynicism regarding efforts to bring about change or improvement are common features of an unhealthy work culture.

To address these issues, it has been long recognized that interventions designed to foster a "culture of safety" within an organization can go a very long way toward reducing the conditions that lead to a high rate of errors and failures. Some of the elements of a safe organizational culture are blamelessness, openness, mutual trust, a commitment to continuous improvement, a belief in quantitative measures and analytics to guide decisions, a shared sense of unity around a higher purpose, and a deep and shared commitment to continuous learning and process improvement, fairness and justice, and forgiveness when human errors inevitably occur. The safety culture in turn leads to enhanced detection and analysis of errors and near-miss events, fostering learning at both the individual and organizational level, leading to progressive performance improvement. Changing an organization's working culture for the better is at the very core of the quality and safety movement in American healthcare.

The classic Russian author Leo Tolstoy once observed (in his novel *Anna Karenina*) that "Happy families are all alike; [but] every unhappy family is unhappy in its own way." Similarly, high-functioning (high-reliability) organizations are all strikingly similar, while there are myriad ways in which an organization can sink into operational and cultural dysfunctionality. The US Agency for Healthcare Research and Quality (AHRQ) defines "high-reliability organizations" as those that can operate in high-hazard domains for extended periods of time without serious accidents or catastrophic failures, and there is a great deal we can all learn from them.

The best-described and classic example of a high-reliability organization is the military aircraft carrier. Despite high "output" and performance pressures, with fighter jets taking off and landing from the small flight deck every 48 to 60 seconds in constantly changing conditions, fatal or catastrophic accidents almost never occur. Like Tolstoy's happy family, all such high-reliability organizations have been described as exhibiting a consistent handful of cultural traits, such as a reluctance to simplify the understanding of complex work processes, a high level of awareness of operational conditions, a deference to the expertise of front-line workers, a commitment to resilience, and a certain preoccupation with the possibility of failure and mindfulness of how things could go wrong (rather than growing complacency since errors and serious accidents are so rare). Blamelessness leads to much better discovery of risks and analysis of near-misses in such a culture, and those near-misses are seen as opportunities to learn and improve. Such systems routinely outperform the expectations of those who rely on them, whereas a dysfunctional organizational culture will inevitably underperform the expectations of those who rely on it—with the predictable result of high and increasing rates of error, an impaired ability to detect error promptly for the prevention of patient harms, and thus poorer overall outcomes.

Creating Cultural Change

Bringing about cultural change in a healthcare organization is a daunting task, generally requiring years of concerted effort on the part of leaders and committed team members. Change of any kind is always difficult for people, even when everyone in an organization agrees that a change of some kind is desperately needed. Sometimes change is not possible without staff turnover, as there are those who would obstruct needed changes or who are unable or unwilling to embrace needed cultural change and adapt to it. A deep literature on organizational psychology and "change management" has been developed, and one can find practical tools, detailed strategies, and methods to enable leaders to accomplish needed cultural change in all types and sizes of organizations. A detailed discussion of these methods is beyond our scope here; suffice it to say that, while having a dysfunctional organizational culture is often a substantial problem, leading to a high rate of systems errors, this type of problem is eminently remediable—albeit only with patience, determination, and long-term consistent effort. Bringing about needed shifts in organizational and leadership thinking and behavior so that cultural transformation can develop "organically" is often the stated goal of expensive consultants hired by healthcare organizations to help deal with these sorts of problems. Uncovering the hidden reasons that underlie people's resistance to change in an organization is an essential first step in the improvement of organizational culture.

Blamelessness

In dealing with the risk of error, a key cultural obstacle in physician groups and healthcare organizations that must typically be overcome first is the tendency to assign blame to individuals and even to identify a "scapegoat" (i.e., a person or persons to punish when harmful errors inevitably occur). I believe that this approach is deeply intellectually dishonest. It is generally subscribed to by leaders who lack a basic understanding of the fundamental nature of human error.

Blaming and scapegoating may serve to reduce short-term psychological tensions that develop within organizations when errors and harms are detected, but they provide only a false sense of security, since most examples of human error are unavoidable and punishing "bad actors" generally will not do anything to reduce the future risk of error.

On the contrary, as we touched upon in Chapter 3, the "blameless culture" and "just culture" models of responding to error—wherein the limits of human fallibility are understood and accounted for, and where all errors and near-misses are able to be brought to light and analyzed without any individual fear of recrimination—have been shown to be extremely effective in error reduction and harm prevention across many industries. It is this realization that prompted Dr. Lucien Leape, a professor at Harvard's School of Public Health and an early pioneer in the field of healthcare quality and safety, to testify before Congress in 2009, saying: "The single greatest impediment to error prevention in the medical industry is that we punish people for making mistakes."

FAULTY TECHNOLOGIES

The technologically based systems (e.g., equipment, computer software) that are used in the healthcare setting are generally designed by people who do not fully understand the circumstances and situations in which they will be used. This often leads to a mismatch between the technology and the people who use it to provide healthcare. Nonintuitive or inconsistent functions of equipment are a well-described source of errors. A simple example in radiology is when there are variable displays of the current and prior images on multiple screens, leading to confusion as to which images are the current ones and which are from the prior comparison study;

errors of this type can lead the radiologist, for example, to erroneously conclude that an abnormality is getting smaller when it is actually getting larger over time. Similarly, inconsistent, incorrect, or difficult-to-find information in the electronic medical record can confuse the history and can lead to diagnostic error or even exposure of patients to contrast agents they may be allergic to. Many of us have experience with voice-recognition systems that can misinterpret speech and that substitute reasonable-sounding and grammatically correct—but clinically incorrect—phrases into reports that in some cases convey the opposite of the intended meaning. While sometimes amusing (a "water-soluble enema" was once transcribed by such as system as a "motorcycle enema"), in many instances the errors are difficult for the radiologist to detect while proofreading. It's even more difficult or even impossible for the reader to realize that the report is not what was intended by the dictating radiologist, leading to miscommunication and diagnostic error. In one recently published study, for example, a full 78% of radiology reports that later required corrections/addendums were due to simple typographical errors!

For this class of error, the root cause is in the interface of people and machines; accordingly, the solutions will generally lie in human-factors engineering, an approach by which machines and software are *adapted to the people who use them* rather than expecting the operators to learn multiple different technological systems and to adapt to each of their quirks. In human-factors engineering, the possible failure modes from people's interactions with devices and software are carefully considered in the design of human–machine interfaces in order to make them more intuitive, predictable, and error-correcting.

FAULTY ERGONOMICS

Clearly the risk of error is increased if the radiologist's working conditions are not optimized. Good ergonomics of the workplace, the minimization of distractions and physical discomforts, avoidance of eyestrain by virtue of optimized image displays and lighting, and even the incorporation of frequent breaks will all likely improve radiologists' performance and accuracy. Working in a noisy, uncomfortable room with inappropriate lighting, suboptimal arrangement of equipment and displays, and frequent interruptions and distractions—a perfect description of the workplaces of most radiologists today—is a formula for enhancing the risk of errors of all types. Intervention strategies based on applying knowledge derived from the science of ergonomics are therefore likely to also have a benefit for radiologists.

Similarly, the use of checklists, structured reporting, streamlined access to information contained within the electronic health record, electronic alerts, and electronic clinical decision support all can play a role toward error reduction. The computerized picture archiving and communication (PACS) systems that are now at the center of the radiologist's work can also be optimized to make it impossible, for example, for a radiologist to display one case while dictating a report on another. Dictation systems can flag "right/left" inconsistencies in dictated reports, thus alerting the radiologist to prevent this common type of reporting error. Electronic databases can help to keep track of clinical reminders, enhance the ease of communication of critical results, and prevent clinical recommendations from being lost to follow-up.

FLAWED COMMUNICATION

The Irish author, playwright, and Nobel Laureate George Bernard Shaw (1856–1950) famously opined that the most common problem with communication is "the illusion that it has taken place." An excellent example, unfortunately, is the radiologist's written report, which is intended to communicate the radiologist's findings and impressions to the physician who referred the patient. As was mentioned in Chapter 2, the formal, written report has, in recent years, become the prime means by which radiologists communicate their impressions to the referring physicians, who rarely opt to receive results in person or by phone, as in the past. Misunderstandings abound in this space, some caused by errors of transcription (now done primarily by electronic "natural language processing" systems rather than by human transcriptionists, as noted earlier) and others caused by a lack of clarity in the language or structure used by the radiologists writing their reports. Confusing or unclear writing, which leads to reports that are not fully or correctly understood by their intended audience, is a large source of systems error in radiology. As noted previously, a particular weakness in this regard is when radiologists are uncertain about a finding and attempt to convey their own level of uncertainty as to what the observed findings may actually mean. Studies have shown that nuances of meaning intended by radiologists when writing their reports are very seldom understood by the physicians who later read those reports.

Providing better training and feedback to radiologists on report writing would be a reasonable strategy to address this problem. One underappreciated aspect of the problem stems from the fact that *spoken language* is different in key ways from *written language*. Radiologists generally dictate their reports—that is to say, they speak them into existence, which requires a bit of a mental shift: they must learn to *speak* in the form of *written* language!

As was noted in Chapter 2, at our center we have tried to address this problem educationally by providing our radiology trainees with an annual report-writing workshop aimed at improving the clarity of their reports. In our workshop we stress that writing a clear, understandable, and succinct report is a separate skill, distinct from the other tasks of diagnostic radiology, and thus requires dedicated effort and practice to master. We show the residents that any piece of writing can be improved by better organization, more appropriate vocabulary, use of good grammar, and simplified sentence structure. The residents spend hours revising and improving actual reports, and their work is discussed in an open forum, with immediate feedback. In our evaluations of the results of this effort, we determined that focused training in report writing was effective at improving report clarity for our residents. I remain convinced that any and all efforts spent by radiologists toward crafting a clearer, better-written report will pay off with reduced errors resulting from miscommunication.

Of course, communication failures are also common in the other direction—that is, when clinicians fail to effectively communicate with radiologists. Published studies have extensively documented systematic failures to provide clear and correct clinical information to radiologists or to correctly frame the clinical question that prompted imaging. While it is both customary and required by many insurers that the clinical reasons for each radiological study are provided, this information is often very cursory and is often incorrect or incomplete. Since physicians are rushed and harried, the clinical comments on radiology requisitions may actually be typed into the requisition on their behalf by others, such as nurses or clerks, based on verbal comments from the doctor, or sometimes with mindless, repetitive use of standard phrases regardless of the true clinical question.

There may be propagation of errors in cases where cutting/pasting of prior study indications leads to a perpetuation of incorrect data being carried forward. In fact, the clinical information provided to radiologists to justify an imaging study is sometimes very misleading, as clinicians or their agents may try to "game the system" by offering an indication that they know will be acceptable for an insurer when in fact they are considering a very different set of diagnostic possibilities. An example would be when "rule out pneumonia" is entered into the system repeatedly and indiscriminately to justify every chest x-ray request that is made, even for cases where there is no reason to suspect pneumonia and instead there may be a strong clinical reason to suspect a different diagnosis, such as congestive heart failure or pneumothorax. Sometimes the simple matter of a doctor making an incorrect choice of which imaging study should be obtained for the clinical diagnostic question at hand can—in and of itself—be a form of communication failure by means of misdirecting the radiologist's focus.

As Dr. Robert Wachter eloquently discussed in his book *The Digital Doctor*, the pervasive adoption of electronic means of medical imaging using PACS, which more or less completely replaced the prior technology of viewing images on film backlit from a viewbox in the early years of the current century, also ended the prior common practice of clinicians' making their daily "radiology rounds," in which essentially all clinical doctors would physically venture to the radiology department to view their films and have an active face-to-face conversation with a radiologist. This intense, personal level of two-way communication was quite effective, as the give-and-take of an in-person conversation, including all of its verbal and nonverbal cues, served as a bulwark against misunderstanding and miscommunication. In current practice, now that clinicians are able to view images remotely from their offices and clinics, allowing them to work faster and serve more patients each day, this type of face-to-face interaction rarely occurs; it is viewed as too time-inefficient. Over recent years, radiologists have also been placed under increasing pressure to work faster and more efficiently and now generally tend to view physician visits and phone calls to discuss cases as a nuisance and a needless interruption as they toil to generate their written reports as quickly as possible, often under time pressure to achieve target measures of report turnaround time (RTAT). As a result, neither party fully benefits from gaining a richer understanding of what the other is thinking.

A broad range of systems-focused management interventions aimed at changing this state of affairs would, by necessity, trade off throughput and efficiency of both parties in order to gain more reliable performance, improved communication, and error reduction—and thus would have both manpower and financial repercussions. It would appear that, from the perspective of those who manage clinical physician practices, the trend toward continuously accelerating the pace of work in the pursuit of predefined revenue targets, with tacit acceptance of increasing risks of provider burnout, communication failures between caregivers, and other types of errors, will most likely proceed unabated unless there is a true paradigm shift in the theory of clinical practice management.

Errors of How We Think

As we noted in previous chapters, over 40 distinct cognitive biases have been described, of which a handful are thought to commonly plague medical diagnosis, including the work of radiologists. These biases can each be thought of as examples of *failed heuristics*.

Heuristics are mental shortcut processes that are developed by training and experience, although some may have been "hardwired" by evolution. Heuristics enable people to make decisions quickly, especially when they are faced with incomplete information or complexities that would normally require a lot of time and mental effort to solve or decide. They are usually adaptive in that individuals must often make quick decisions with incomplete information. From an evolutionary standpoint, one can easily understand how having a good set of heuristics might provide a significant survival advantage, both from the standpoint of efficiently finding food and other needed resources and avoiding perils in a complex environment.

An important part of a person gaining expertise is the development of heuristics, since they allow experts to more efficiently carry out discriminatory tasks in their domain of expertise. For radiology, this means that some patterns of findings on patient images are simply and quickly recognized, allowing rapid decision making as to the final diagnosis. And, as for most domains of expertise, developing a good, strong set of radiology heuristics makes radiologists more effective and efficient at meeting the challenges of their work, often allowing them to bypass the slower, more effortful and cognitively demanding approach of careful, deliberate, and thoughtful analysis of every case. This is the essence of the "dual process" theory of decision making, a concept originally developed in the 1970s and for which Amos Tversky and Daniel Kahneman were awarded the Nobel Prize in Economics in 2002.

Heuristics are very often correct, and they can be immensely helpful in meeting life's challenges, but they also can go wrong, and sometimes do so in predictable ways. Here is an example taken directly from Tversky and Kahneman's original paper, published in the journal *Science* in 1974:

> For a period of one year, each hospital recorded the days on which more than 60% of the babies born were boys. Which hospital do you think recorded more such days:
>
> - The larger hospital (21)
> - The smaller hospital (21)
> - About the same (that is, within 5% of each other) (53)
>
> The values in parenthesis are the number of undergraduate students who chose each answer.

If you found yourself agreeing with the majority of Tversky and Kahneman's undergraduates (i.e., that the birth ratio of babies in each of two different-sized hospitals would be about the same), you likely reached your answer by using heuristics—you were quickly able to conclude that the chance that a newborn baby will be either a boy or a girl in a normal population is described by the same baseline probability in both large and small hospitals. You probably arrived at this conclusion quickly and fairly effortlessly, and with great confidence. But you would actually be incorrect: the answer is that the smaller hospital is the one that is *more likely to deviate* from the baseline probability of about 50% to 51% of newborns being boys, due to the smaller sample size! If you reached the correct conclusion, you either have some expertise in statistical reasoning or else you took a little more time and effort to think the question through before arriving at your (correct) conclusion.

Kahneman and Tversky described heuristic-based thought as "type 1 thinking" and the more laborious, time-consuming, and effortful type of reasoning as "type 2 thinking." It is believed that expert radiologists are able to increasingly rely on their type 1 thought processes, and typically do so with great accuracy. This is a result of their years of training and experience, in which they informed their heuristic by means of considerable repetition to allow accurate pattern recognition. This is the reason experienced radiologists are able to interpret a much larger number of studies in a given amount of time than are novices, and do so with similar or even higher accuracy, despite working much faster. But these same expert radiologists will switch to the much slower type 2 (deliberate, methodical) thought processes when an image pattern is not recognized or when they encounter confusing or confounding aspects. This is sometimes thought of as a "toggle function," whereby most of the expert's time is spent in type 1 thinking but he or she will "toggle" back and forth to type 2 thinking when needed. For any identifiable type of task or case, however, with repetitive type 2 thinking experiences, experts will eventually inform their heuristic and be able to approach future examples using type 1 (easier, faster) thinking. Croskerry and others have termed the tendency to move as quickly as possible away from type 2 and toward type 1 thought processes to be the "cognitive miser" function. It is always at work but may be particularly miserly at times of fatigue (especially cognitive fatigue), cognitive overloading, and sleep deprivation, which can perhaps in part explain why people are more prone to errors under these conditions.

So type 1 thinking is a hallmark of expertise, but when heuristics lead us to the wrong conclusion, we consider them to be cognitive biases, the root cause of a large class of errors.

We discussed the three cognitive biases that most commonly lead to misdiagnosis among physicians in Chapter 2: (1) *anchoring bias*, in which one's initial impression becomes "stuck," despite subsequent discovery of contrary information; (2) *confirmation bias*, the tendency to seek confirmatory evidence to support one's hypothesis while ignoring or discounting any contrary evidence that is found; and (3) *availability bias*, in which a diagnosis is considered to be more likely because the radiologist has recently seen a case.

The principal strategy for combatting these types of errors (i.e., errors based in the way we think) centers around the idea of *cognitive de-biasing*—and the greatest impact of this approach for error reduction can be expected if we focus our efforts on the "top three" biases. A handful of cognitive de-biasing approaches have been advocated and tested (de-biasing is the subject of a rich literature), and most involve the need for "metacognition"—that is, thinking about one's own thinking. This approach involves having a heightened awareness of when a bias may be present, reflecting on one's own thought processes, and mentally asking oneself questions aimed to counteract the effects of anchoring, confirmation, and availability biases, such as "What else might this be? And where else/how else might this be?" The results of this approach have shown some promise in the medical literature for physicians arriving at complex diagnoses (see, for example, the papers by Drs. Pat Croskerry and George Bordage). It seems intuitively obvious that this approach would work for radiology; however, for radiologists, the success of applying cognitive de-biasing approaches has yet to be fully demonstrated experimentally.

Perhaps the approach to error reduction that has received the greatest effort over the years in radiology has been educational, based on the *knowledge gap hypothesis*—in other words, that errors largely result from gaps in the knowledge of individual radiologists, who will "see more if they come to know more." This is an extension of the famous adage from the philosopher Johann Wolfgang von Goethe (1749–1832), who famously opined, "One sees what one knows." The intervention of choice for this approach is continuing medical education (CME). Most practicing radiologists accrue 50 to 100 hours of CME each year, much of it reviewing material they already know to varying degrees, with the hope of "closing their knowledge gaps" for both unusual and commonly encountered clinical scenarios. By contrast, most radiologists accrue very few CME hours that are focused on learning about new discoveries or experimental methods under development. When one peruses the titles of the myriad offerings of CME classes and meetings, printed CME articles for radiologist learning, online teaching modules, video lectures, and the like, it is striking how often one encounters the word "pitfall." CME offerings seem primarily designed to help learners avoid one or another identifiable "pitfalls," or knowledge gaps, that will lead them directly to making a diagnostic error. From a cursory review of the priorities and titles of the available body of CME, one might reasonably conclude that the practice of radiology primarily consists of the avoidance of one or more pitfalls, which threaten each practicing radiologist (repeatedly and in rapid sequence) throughout each working day!

While the knowledge gap hypothesis is clearly not entirely without merit, as unknown diagnostic possibilities will never be considered, it is also clear that over 60 years of intensive participation in well-designed and well-presented CME has done little to reduce the overall prevalence of errors in radiology. This is because only a very small fraction of radiology error can be truly attributed to gaps in radiologists' knowledge. In the study by Kim and Mansfield that we discussed previously, a large number of radiologist errors were categorized and classified. In this very large, multiyear sample, errors of incomplete knowledge accounted for only 3%.

But perhaps part of the overall failure of CME to prevent errors may be related to the observation that CME content itself may be biased toward illustrating case examples that are considered "classic" for various disease states rather than exposing learners to the full range of experience and to the much broader spectrum of how any given disease may be manifested on various types of images. In this way, past errors, knowledge-based biases, and misinformation can be propagated from teacher to learner and from generation to generation, cementing faulty reasoning and misinforming experts' heuristics.

Errors of How We Are Made

The premier example of an error type whose root cause appears to be based in our shared human biology is the common "perceptual error," as was illustrated in Figure 2.1 in

Chapter 2. As we discussed in that chapter, this type of error generally involves radiologists simply not seeing (or at least not *recognizing or remembering* that they have seen) an abnormal finding that is deemed to be sufficiently conspicuous in retrospect, and sometimes even seems obvious to both the original observer and others. This is the most common type of radiology error by far, as we have discussed, accounting for almost two times the error rate of the next most common error type that radiologists are prone to make (satisfaction of search). Both, I think, are examples of errors derived from the way the human visual-perceptual system is designed and functions—that is, errors of *how we are made*. By prevalence, this category comprises the two most important types of error that radiologists face. These errors are the most poorly understood, and effective intervention strategies are most severely lacking for them.

It has been reliably demonstrated in extensive research carried out over the past several decades that there appears to be an impenetrable "base rate" of these perceptual errors, known to occur on the order of 3% to 4% consistently across all subdisciplines of radiology and across all radiologist working situations. The occurrence of any individual error of perception is generally unpredictable; these errors are not limited to only subtle findings or to a certain class of abnormality or type of imaging study. Their occurrence is seemingly random, in that multiple observers will tend to have differing missed findings, though each observer will have essentially the same rate of misses, for any given sample of radiological interpretations. The pervasiveness of this phenomenon strongly suggests that its origin is biological.

DOUBLE-READING

The only intervention that has ever been effective to mitigate this basic and common type of perceptual error is *double-reading,* in which radiological studies are interpreted twice, once each by two radiologists, ideally with the second interpreter unbiased by the impression of the first interpreter. The rationale for double-reading is that if the first reader has a "miss" rate of 3% (0.03) and a second reader also has a miss rate of 3% (0.03), and that these misses are independent of each other (and are often unrelated to the subtlety or difficulty of the interpretation or to the knowledge base of the first reader), and thus that each reader's miss is a random event, the combined miss rate from double-reading by two independent readers will, probabilistically, amount to the product of their miss rates, or 0.0009 (less than 1/10th of 1%). This is a very significant performance improvement!

Accordingly, double-reading has been advocated since the time of Garland's first description of radiological error in 1949. The advantage is based on the straightforward presumption that two visual-perceptual systems are better than one! All of the available data collected from 1949 to the present seem to strongly support this conclusion. It also seems as if it should be an extremely easy and straightforward intervention to make: simply arrange to have every imaging study separately interpreted by two radiologists. However, the sheer number of radiological examinations requiring interpretation and the relatively low number of trained radiologists available to interpret them would not be sufficient to support double-reading of every imaging study, and the costs involved would be so high as to be widely thought to be prohibitive.

With one key exception. In academic centers, in the training model for new radiologists, all after-hours imaging studies (including x-rays, CT, Ultrasound and MRI scans, etc.) are generally initially interpreted by a radiology resident or fellow (a fellow in radiology is a trainee with a higher level of experience) overnight, and then the study is subsequently "over-read" (double-read) the following morning, often several hours later, by a teaching attending radiologist, who then provides instruction and feedback to the trainee. This has been the prevailing workflow model for academic radiology since the first radiology residencies were created in the 1930s. Typically, residents would provide radiologist service coverage to the hospital's emergency department at night, functioning autonomously in this role, without an attending radiologist present, and subsequently would have their work double-read the following morning by the appropriate subspecialist attending radiologist (e.g., the bone studies would be over-read by a musculoskeletal radiology subspecialist, the brain studies would be over-read by a neuroradiologist, and so forth). This provides the double-reading and subspecialist expertise benefit at a very low cost and has worked well for academic radiology for more than eight decades. In recent years, however, this overnight resident coverage model has come under pressure due from key groups of customers for radiology services, particularly emergency department physicians, who prefer to have an attending radiologist provide them with final interpretations on a 24-hour basis. While substituting an attending radiologist for a resident at night *seems* to be a move toward a higher standard for quality of care, in fact, the data suggest that it is not. As it happens, the error rate for residents interpreting after hours autonomously is quite comparable to that of their supervising attending radiologists and in most studies is actually somewhat better (on the order of 1% to 2%) than attending radiologists interpreting similar studies under similar conditions by themselves (typically the attending radiologists will have a 3% to 4% error rate). The resident model has the added benefit of a double-read by a "fresh" subspecialist the following morning (i.e., someone who is just starting his or her workday) who has subspecialist expertise in each type of case.

Because of cost concerns, the double-reading step that is part of the resident/attending dyad will pretty much always be omitted for a "solo" overnight attending's interpretations, as attending-level reports are considered "final." Based on the available data on error discovery and prevention, however, this loss represents a significant degradation in the quality of service provided, since the error-detection function of the double-reading is so very powerful.

The power of double-reading was again demonstrated in a very compelling recent study by Balthazar et al.: in their study, it was found that radiology reports that were generated with one reader only (i.e., without a trainee involved) were *12 times more likely* to require a later formal amendment or addendum (an indication that the original report contained one or more errors) than were reports in which two readers (i.e., a trainee and a supervising radiologist) both contributed to the final report. Since we can assume that not all errors were discovered and thus not all erroneous reports were corrected, this study's result only allows us to make a *lower-range estimate* of the *minimum* value conferred by double-reading: it provides *at least a 12-fold benefit* for the reduction of error.

A NEUROBIOLOGICAL HYPOTHESIS FOR PERCEPTUAL ERROR

But what is the underlying neurobiological mechanism by which these perceptual errors (i.e., the main type of errors by prevalence in radiology) occur? The answer remains unknown, of course, but an attractive new hypothesis, which is the focus of current study in our laboratory at Penn State, in collaboration with scientists at Johns Hopkins and Emory universities, draws from the observation that within the human brain there are functional networks of connected brain areas that work in concert under certain situations. For example, there is a brain network that operates when one is at rest, or not engaged in a mental task, the "default mode network (DMN)," and there are other networks of brain connections, involving different cortical brain regions, that are activated only when one has a task to perform that requires concentration, such as the "fronto-temporo-parietal (FTP) network." In general, the DMN is involved with inwardly directed mental activity and the FTP in externally directed mental activity (a bit of an oversimplification, but generally true).

So when one of these brain networks is engaged, the other is usually suppressed, so that the brain essentially toggles between them. We can observe these phenomena easily using fMRI brain imaging, as illustrated in Figure 4.4 of the prior chapter. At rest, the DMN is active, but when one begins a task requiring attentional focus, the FTP is activated and the DMN is suppressed. When the task is completed, the activation pattern reverses, and so forth. All well and good. But we also sometimes see unpredictable, random, episodic, and spontaneous events where the DMN is reactivated during the performance of an active mental task, such that there is *simultaneous co-activation of these two brain networks,* which usually are seen to work in opposition to one another. It is unclear why this would happen, but it has been shown that when it does, it can result in unpredictable episodes of involuntary attentional lapses—known in the psychology literature as "task-unrelated thought" or, more commonly, "mind-wandering." It is thought that the DMN essentially coopts the resources of the attentional brain network at these moments, redirecting the mind toward "internal," task-unrelated thoughts rather than "external," task-related thoughts.

Careful experimentation involving the evaluation of brain activation states and behavior in human subjects has shown that everyone is susceptible to these attentional lapses and that we are very rarely, if ever, aware of them when they occur. Further, recent studies using fMRI brain imaging have shown that when the DMN is activated simultaneously along with task-oriented brain networks, this can result in a phenomenon known as "perceptual decoupling," essentially a brief episode where a person is unresponsive to any outside perceptual stimuli—and generally without any conscious realization. The phenomenon has been best studied in patients with attentional disorders, such as attention-deficit/hyperactivity disorder, but is now believed to occur universally in the brains of all human beings, including radiologists. It is very tempting to believe that this phenomenon, or something like it, is happening when radiologists—randomly and inexplicably—fail to see something on an image that they later appreciate as being obvious to them in retrospect.

This novel neurocognitive hypothesis for the underlying cause of radiologists' basic perceptual errors suggests that it may be possible to improve their performance using interventions designed to detect this mind-wandering when it occurs and to promptly alert radiologists that their image search may be compromised. The role of cognitive loading and cognitive fatigue is also worth considering—it is unclear whether these factors, which can be modulated by systems changes, will impact the brain-level phenomena that are hypothesized to underpin the most common "biologically based" type of error.

Of course, the remedy for common perceptual errors cannot involve having all radiology interpretations performed by radiologists lying within an MRI unit while their brain network patterns of activity are actively monitored in real time! Something more practical will need to be found. Assuming a practical, unobtrusive means to detect brain-based attentional lapses can be created, it still remains to be seen whether such an approach will be effective toward reducing the main cause of radiologist error, but this is our laboratory's current research goal. In theory, such a neurocognitive alert system could also be used to warn a radiologist when "cognitively dangerous" conditions are present, such as excessive mental fatigue or "cognitive depletion," at least if our ongoing research eventually shows that such factors actually matter in terms of the risk of perceptual error. Mindfulness practice, a type of mental discipline that can be learned with practice, has also been shown to reduce the incidence of mind-wandering.

But until a practically workable neurocognitive error/risk-detection tool can be created, the only intervention that has ever been shown to be effective for the basic perceptual error, as discussed in some detail earlier, is still *double-reading*. Recent research studies, one published as recently as April 2018, have shown a benefit to double-reading of mammograms, with reduction in both false-negatives (i.e., misses) and false-positive results. But, as noted earlier, double-reading has never been widely adopted because of the high cost and realistic constraints on the manpower needed to accomplish it universally.

One suggestion has been to apply double-reading selectively, perhaps only to screening mammography or to other high-risk radiological examinations. This approach might be logistically feasible, but it still leaves the vast majority of imaging studies (outside of training hospitals, at least) without the benefit of a second reader.

SATISFACTION OF SEARCH

Similarly, the phenomenon of satisfaction of search, which is the second most important error type in radiology by prevalence, may actually be better understood in terms of a neurocognitive mechanism (i.e., a fundamental neurobiological process), resulting from a visual perception feature known as *salience.* Recall that this type of error is thought to be involved in explaining the lower probability for a radiologist to identify a second or third abnormality present on a radiological study after a first abnormality is identified. Often it is NOT the more subtle lesion that is missed! Terming this type of error "satisfaction of search" implies a mechanism—that is, that finding an abnormality is psychologically satisfying in some way, leading to premature closure of the search, akin to Dr. Herbert Simon's Nobel Prize–winning description of the human trait of "satisficing" in economics (i.e., the tendency to terminate one's effort once a satisfactory or "good enough" result is achieved). This is essentially the opposite of the tendency toward "optimizing," wherein only the very best final solution will do. However, eye-tracking and time studies of radiologists do not reflect a curtailed search pattern nor a shortened search time in cases where this type of error is detected, so another mechanism must be at work. One attractive hypothesis has to do with the visual-perceptual concept of *saliency*, whereby features on an image that have high meaning to the viewer (e.g., a lung nodule on a chest x-ray or a cluster of calcifications on a mammogram) tend to capture and hold the viewers' attention, as we touched on in Chapter 4.

By this reasoning, the presence of a high-saliency finding on an image can capture viewers' attention in such a manner as to prevent them from noticing the (sometimes equally salient) findings in other areas of the image, especially if the two findings are fairly near each other. This hypothesis is currently being tested in active research (I participated as a subject in such a recent study, in the Radiological Society of North America's visual perception laboratory, set up during the annual meeting in Chicago in 2017) that compares the risk of satisfaction-of-search errors on planar imaging, where the two findings are seen in the same visual field, versus tomographic imaging, where the "competing" findings are more visually separated. This work is ongoing.

COMPUTER-AIDED DETECTION

The new frontier for radiology is, of course, the burgeoning field of machine learning and "artificial neural networks," sometimes known as "artificial intelligence" (AI). There has been quite a bit of hyperbole on this subject. Clarifying the current and likely future state of AI as applied to radiology is the focus of the final chapter of this book. Computational methods for extending the vision and reach of radiology do indeed hold promise for error reduction, although it is widely believed that human visual observers, and their biologically based means of perception, will remain the mainstay of diagnosis for the foreseeable future. As was noted in a prior chapter, past efforts to use computers to aid diagnosis, known as "computer-aided diagnosis" (CAD), have largely failed in the past, primarily owing to an unacceptably high rate of false positives, and the concept of CAD has fallen into low repute among radiologists. However, this could change as computational methods improve, particularly with "deep learning" algorithms, in which computers essentially teach themselves what to "look" for in an image. We will focus on this topic in greater depth in Chapter 6.

Information technologies based on eye-tracking have been used to analyze radiologists' search patterns and in the future might be applied for the purpose of error reduction by providing immediate feedback to radiologists on areas they may have neglected on their search. For instance, a "map" could be provided showing what the radiologist has looked at (or not) on an image. The radiologist can view this before signing off a case. Or technology could actively guide the radiologist's search in real time, perhaps by altering the brightness of regions of the image that the radiologist has neglected, a technique known as gaze guidance. This can be applied subtly at the point of care and thus may be used unobtrusively in the clinical setting. You may recall that in prior chapters we discussed how eye-tracking studies have shown that radiologists often fixate their gaze for some time on areas of images where positive findings are later missed, without being consciously aware that they have done so. One idea to use this knowledge is to use eye-tracking to create a colorized "heat map" overlay or another annotation to the image that highlights areas where the radiologist's vision lingered during the search; areas of high visual dwell time may suggest that a consciously unrecognized abnormality may in fact be present. The eye-tracking and computer analysis technology could thus be used to "nudge" the radiologist to look again—more deliberately—at these areas, and possibly avoid making an error.

ROADMAP FOR THE FUTURE: REDUCING ERROR IN RADIOLOGY

In summary, we are faced with an array of potential strategies to reduce radiologists' errors in practice. Some have failed in the past, others have had only partial success, and still others are unproven or have not yet been tried. Perhaps the best approach is to simultaneously apply a combination of these strategies, such as (1) for each radiologist to maintain a state of constant vigilance in interpretation and (2) a healthy degree of skepticism regarding his or her favored diagnoses, as well as (3) a conscious awareness of when heuristics are being used and how they might fail, (4) frequent focused metacognition to combat cognitive biases, which are nearly always present, by asking de-biasing questions, such as "What else might this be? Where else/how else might this be?," as well as (5) a full range of systems interventions to optimize the work climate and culture, as we've discussed in some detail, as well as efforts to reduce or eliminate distractions, interruptions, and variability in practice at all levels, (6) a program of continuous lifelong learning for every radiologist to close knowledge gaps, and (7) a consistent focus on radiologists to be clear and effective in their communication with other healthcare providers and with their patients, especially as regards written communication, so that their message is not lost or misunderstood.

Radiologists would also clearly benefit from the development of systems that can reliably provide them with accurate performance feedback, which they can use to improve. Sir William Osler, the architect of modern medical education, once opined that there are only two types of physicians: those who are overconfident and those who are underconfident! But in the absence of meaningful (and sufficiently rapid) performance feedback, it is a natural human tendency to become increasingly overconfident in our abilities, and thus to perpetuate errors. In many fields, rapid and reliable feedback is the principal way that experts improve their own performance, and accurate, unbiased feedback is the best antidote for both overconfidence and underconfidence in our own abilities.

In radiology, a valuable source of such feedback may come directly from patients, who are increasingly reading radiologists' reports via online portals. In recent years many radiologists, myself included, have become accustomed to being contacted directly by patients who read our reports online, promptly addressing any errors that they have detected, especially to correct information from their medical history that we have mentioned that was incorrect or misleading—not to mention catching all of our grammatical, spelling, and typographical errors!

Future research, including our own, will hopefully extend the list of useful approaches for error reduction to include measures designed to remedy neurocognitive factors in attentional focus and perceptual discontinuities, factors that seem to be part of our basic biology but that may underlie the most common types of error to which radiologists are prone.

NOTE

1. Here is a link to an excellent online tutorial on Lean/Six-Sigma, which is offered free of charge from the Massachusetts Institute of Technology (MIT Open Courseware): https://ocw.mit.edu/courses/aeronautics-and-astronautics/16-660j-introduction-to-lean-six-sigma-methods-january-iap-2012/lecture-videos/.

FURTHER READING

AHRQ Patient Safety Network Primer on High Reliability. Last updated November 2017, available at: https://psnet.ahrq.gov/primers/primer/31/high-reliability

Arenson RL. Factors affecting interpretive accuracy: how can we reduce errors? *Radiology* 2018; 287(1):213–214. https://doi.org/10.1148/radiol.2017172781

Bordage G. Why did I miss the diagnosis? Some cognitive explanations and educational implications. *Acad Med* 1999; 74(10):S138.

Bruno MA, Petscavage-Thomas J, Abujudeh HH. Communicating uncertainty in the radiology report: Commentary. *AJR Am J Roentegenol* 2017; 209(11):1006–1008. doi:10.2214/AJR.17.18271

Bruno MA, Petscavage-Thomas J, Mohr MJ, Bell SK, Brown SD. An open letter: the radiologist's report in the era of the patient web portal. *J Am Coll Radiol* 2014; 11(9):863–867.

Busby LP, Courtier JL, Glastonbury CM. Bias in radiology: the how and why of misses and misinterpretations. *RadioGraphics* 2018; 38(1):236–247. https://doi.org/10.1148/rg.2018170107

Croskerry P, Cosby K, Graber ML, Singh H. *Diagnosis: Interpreting the Shadows*. Boca Raton, FL: CRC Press, 2017. ISBN 9781409432333.

Croskerry P, Singhal G, Marmede S. Cognitive debiasing 1: origins of bias and theory of debiasing. *BMJ Quality & Safety* 2013; 22(Suppl 2):ii58–ii64. doi:10.1136/bmjqs-2012-001712

Croskerry P, Singhal G, Marmede S. Cognitive debiasing 2: impediments to and strategies for change. *BMJ Quality & Safety* 2013; 22(Suppl 2) ii65–ii72. doi:10.1136/bmjqs-2012-001713

Debono DS, Greenfield D, Travaglia JF, et al. Nurses' workarounds in acute healthcare settings: a scoping review. *BMC Health Services Research* 2013; 13:175. http://www.biomedcentral.com/1472-6963/13/175

Deming, W.E. *Out of the Crisis* 2012 Jul. ISBN 9780262541152.

Kegan R, Lahey L. The real reason people won't change. *Harvard Business Review*, November 2001.

Sokolovskaya E, Shinde T, Ruchman RB, et al. The effect of faster reporting speed for imaging studies on the number of misses and interpretation errors: a pilot study. *J Am Coll Radiol* 2015 Jul 1; 12(7):683–688.

Wachter R. *The Digital Doctor: Hope, Hype, and Harm at the Dawn of Medicine's Computer Age*. New York: McGraw Hill, 2015. ISBN: 978-0-07-184946-3.

6

ARTIFICIAL INTELLIGENCE AND MACHINE LEARNING IN RADIOLOGY

A 24-year-old woman with many chronic health problems and a history of several prior abdominal surgeries is admitted to the hospital to evaluate and treat recurrent, severe abdominal pain. The surgeon suspects small-bowel obstruction, likely the result of adhesions from her multiple prior abdominal surgeries, and explains to the patient and her family that surgery may be needed that very night. A computed tomography (CT) scan is obtained, which is interpreted by a radiologist as showing a "nonspecific" bowel pattern, somewhat abnormal but without clear evidence of bowel obstruction, so the anticipated surgery is not performed; instead, the patient is observed in the hospital for an additional 2 days, during which time her pain persists. Each day an abdominal x-ray is obtained, and each one also shows a "nonspecific bowel gas pattern" in the opinion of the attending radiologist. The surgeon worries that something may have been overlooked, so he chooses to revisit the original CT scan.

He walks over to the radiology department and informally reviews the images with another radiologist, one not previously involved with the case. This second radiologist indeed finds subtle evidence of partial small-bowel obstruction on the images that was not appreciated by the original interpreting radiologist. The anticipated corrective surgery is then scheduled and proceeds on the morning of hospital day 4. The surgeon finds and lyses a number of small-bowel adhesions that were causing the partial obstruction, and the patient recovers uneventfully.

There is a currently a great deal of excitement and anticipation, both among radiologists and within the greater medical imaging industry, about how artificial intelligence (AI) and computer algorithms based on machine learning (ML) and deep learning (DL, a specific form of ML) might profoundly impact diagnostic radiology in the near future. There have been a large number of published articles on these topics, both within the medical literature and in the popular press, and a great many scientific session hours have been devoted to discussing these topics and their implications at the major radiology meetings worldwide in recent years. The American College of Radiology has recently established a Data Science Institute and an AI Advisory Group,[1] and the Radiological Society of North America has even initiated a new scientific journal, *Radiology: Artificial Intelligence*, which is set to launch in early 2019 under the editorial leadership of Dr. Charles Kahn of the University of Pennsylvania. This new journal has been specifically chartered to disseminate new advances in AI research as applied to the field of diagnostic radiology.

There is widespread hope that AI methods, especially those based on artificial neural networks (i.e., computer software designed to mimic the functioning of biological brains) will eliminate diagnostic error and bring about currently unimaginable benefits. There is also a fear that radiology as we have known it will come to an end and that most, if not all, currently practicing radiologists will soon join the ranks of the unemployed. Many experts have come to believe that AI may soon be able to significantly reduce the prevalence of all types of radiologists' interpretive errors, including basic perceptual errors, by providing effective computer-aided detection of findings on images and computer-aided diagnosis (CAD), a goal that has been previously elusive. In this final chapter we will explore these new computer-based methods, explain what they consist of, compare them to what has been tried (and failed) before, and attempt to place AI, including (and especially) the various types of ML and DL, into a broader perspective.

Some of the breathless media coverage of AI, artificial neural networks, ML, and DL is, of course, hyperbole. Many authors have speculated that DL algorithms will soon be able to interpret imaging studies more accurately, faster, and at a fraction of the cost of human radiologists, and thus will soon be putting all radiologists out of work, just as automation eliminated the bulk of manufacturing jobs in recent decades—although even the most advanced DL systems seem very far from being able to do so. The internationally famous AI computer scientist Dr. Geoff Hinton of the University of Toronto (who now also works for Google) opined at an international ML conference that "we should stop training radiologists now."[2] Although he later was reported to have recanted that particular statement, he has also expressed the opinion that, at some point in the future, "probably machines will get smarter than people in almost everything."

A very recent editorial published in the *Journal of the American College of Radiology*, authored by Dr. I. Ray Kirk et al., was entitled "The Triumph of the Machines." Dr. Kirk and his co-authors harkened back to decades of science fiction novels in which AI makes even the most skilled human beings redundant and creates serious ethical dilemmas about what it means to be "human," citing Philip Dick's now-classic novel, *Do Androids Dream of Electric Sheep?* Perhaps Kurt Vonnegut's (1922–2007) debut novel *Player Piano*, which was first published in 1952, is more to the point; in it, Vonnegut envisioned a future world in which the vast majority of people are unemployed, as a result of near-universal automation, such that there is very little need for human labor of any kind except for a few managers and a handful of engineers needed to maintain the equipment. Computer scientist Daniel Wilson's 2011 novel *Robopocalypse* went even further: in it Dr. Wilson envisions a near-future where self-aware AI assumes control of all electronic systems and wages a genocidal war against humankind! The recent success of DeepMind's "Alpha Go" algorithm, which beat a world-champion player in the complex game of Go in 2016, has helped to fuel fears that human beings' mental capacity is being rapidly surpassed and that soon even world-class experts in complex, specialized fields will be easily replaced by AI machines. A DL algorithm very recently developed by researchers at the University of California at Irvine has been able to teach itself how to solve Rubik's cube, a notoriously difficult puzzle to master (see McAleer et al.), and other DL algorithms have allowed computers for some time to recognize spoken words almost as well as a person and have made driverless cars possible, although perhaps not at an acceptable level of reliability.

Algorithms of this type are currently being used to guide key decisions in finance and other sectors, including such decisions as who should receive a loan and who should be let out of prison on parole, albeit always working with human supervision. Driverless cars, another highly anticipated DL application, have been remarkably successful as self-learning algorithms go, but with occasional errors and catastrophic failures, including failure of the supervising human operator who is expected to take over in the event of a system error, with well-publicized tragic consequences and loss of human life.

The current consensus in the field is that, despite Professor Hinton's dire warning, AI won't actually be replacing radiologists any time soon. But it does seem likely that AI, including computerized artificial neural networks, ML and DL, will bring useful tools into existence that can augment radiologists' work and improve their performance in significant ways, including, hopefully, for error reduction.

Some leaders in the field imagine a (very near) future in which ML-based AI tools in radiology will actually be able to "preview" all radiological examinations immediately as they are produced and will be able to "pre-analyze" and "triage" them very rapidly, rank-ordering them for likelihood of positivity before they are ever seen by a radiologist. These algorithms, as imagined, will be able to determine which imaging studies should be prioritized and viewed by a radiologist first in a long list, for example, based on the AI detection of possibly urgent or critical findings. It is anticipated that the computer will also be able to pre-dictate the radiologists' written reports for them, thereby adding efficiency to the radiologists' work.

The algorithms can already do a fairly good job at "segmenting" images, allowing the machine to make accurate and detailed measurements of objects depicted, such as lung tumor volumes, organ dimensions, and the like, even comparing them across numerous prior studies and charting interval changes quantitatively, which are otherwise tedious and time-consuming tasks for human radiologists. And some experts have envisioned a future world in which the work of radiologists will also be continuously monitored by AI algorithms, which can "step in" to correct interpretation errors and can alert radiologists when they have overlooked something on an image, when they may have misstated something on a report, or even if their visual search pattern deviates from their usual one—to suggest that they may be fatigued and need a break—and perhaps, in some cases, can even prod radiologists who seem to be getting distracted or are underproductive in finalizing imaging studies at some predetermined speed. In this way, AI might become sort of a hybrid between a spell-checker in the radiologist's word-processing software and an electronic Orwellian "Big Brother" who is always watching the radiologist and evaluating his or her work performance. One wonders how radiologists will feel about having such an unblinking electronic supervisor watching their every action, one that is itself never distracted or fatigued. Will the new electronic "boss" have no patience for radiologists who are "merely" human, and who make errors?

Some of these concerns are very realistic, of course, while others may be the product of fanciful, paranoid, or even magical thinking! Time will tell. But there are clearly a number of tasks that computers are already able to do better than people can. In order to separate reality from fantasy and gain a better appreciation for the potential impact AI might have for reducing human error in radiology, we must first gain a basic understanding of how these systems work and the current state of the art in AI as applied to medical imaging.

ARTIFICIAL NEURAL NETWORKS, ML, AND DL

In general, AI refers to any technique that enables computers to mimic human intelligence. This would include fairly straightforward computer programs using logic and decision trees and so forth that have been available for decades.

Artificial neural networks are based on the idea that, if computers were designed so that their data structures were to mimic actual biological brains, they would be better able to perform complex "thinking" tasks. To accomplish this, computer logic architectures are created in which there are simple, discrete data-processing elements, each designed to mimic the structure of neurons in biological brains, with individual input units, hidden layers of processing units, and output units, with a large number of interconnections between each node. Each of these individual "perceptron" nodes, the computer equivalent of neurons in a biological brain, is actually a small packet of computer code that functions autonomously in a relatively simple way, known as its "activation function," with each unit receiving inputs from multiple sources, making some sort of simple decision or calculation from these inputs, and then feeding the output forward to myriad other such "artificial neurons" or networked clusters of them. Just as in a biological brain, each of these small units of computer code receives a large number of inputs from other equivalent "neurons." In turn, each produces some sort of "decision," yielding a somewhat refined and reduced number of outputs to the next layer, in a "feed-forward" process.

The basic idea is illustrated in Figure 6.1. The artificial neurons connected to the "input" end receive data (e.g., images, words, voice recordings), perform simple analytical operations on them, and then transmit this "partially digested" information into a deeper analytical layer of artificial neurons, each of which are highly interconnected and seek to make associations between the data. These, in turn,

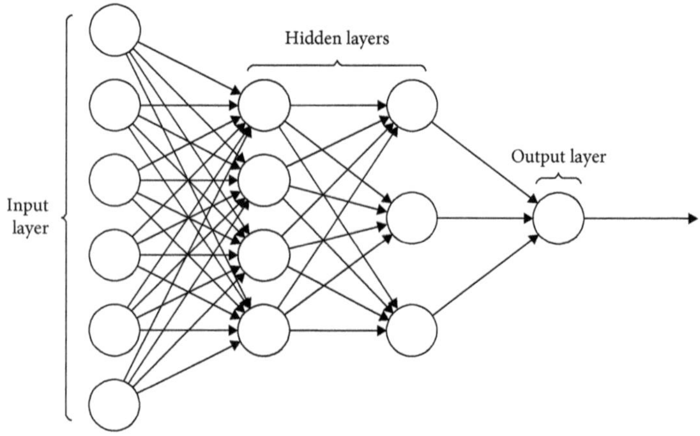

FIGURE 6.1. Basic architecture of a simple artificial neural network, such as that used for ML and DL. The circles represent "perceptrons," fundamental blocks of computer code that perform simple analytical functions, analogous to biological neurons. These basic units are interconnected in a cascading, "feed-forward" fashion through one or more "hidden layers" of analysis and ultimately to an output layer. Interconnections become less numerous as one approaches the output layer. Such a construct is designed to allow a computer to mimic the basic information-processing structure of a biological brain. Image: Public domain.

communicate with a collection of neurons connected to the "output" end, where the user (human, presumably) is able to receive the output (classification, or matching of the words to images or the meaning to sounds).

ML is a subtype of AI that includes any type of computer algorithm capable of allowing the computer to learn new information or "skills" so that it can solve problems without being explicitly programmed to do so. ML includes a large variety of complex statistical/probabilistic calculation elements, especially Bayesian reasoning, decision trees, and statistical learning methods such as support-vector algorithms, that allow the computer to improve at tasks with greater "experience," in an iterative fashion. A computer program is said to learn from experience if its performance at some task improves with experience—that is, with more data.

Supervised machine learning is when the program is "trained" on a predefined set of data, the "training set," that includes an "answer key" for every example, which then forms the basis of its later decisions when given new data. *Unsupervised* machine learning is when the program is simply given a large amount of data and asked to find patterns and relationships within that dataset, without any training examples (no answer key). The algorithm must determine its own criteria for clustering the data. This is the usual approach to "big data science," where datasets are said to be "mined" for unanticipated relationships, such as the impact of various measurable parameters on the likelihood of hospital readmission after discharge.[3]

DL is a subset of ML in which a multilayered neural network architecture is set up and exposed to training data—usually requiring vast amounts of such data—*and then the algorithm trains itself* to perform any desired tasks, establishing for itself its own criteria by which to make discriminations. This is the approach that has been successfully used to develop the now-ubiquitous speech-recognition software and so-called natural language processing, among other useful tasks.

In artificial neural networks of all types, whether used for ML or DL, the layers of computer code are set up in such a way that the network is able to take the data from the "input units," which are visible to the user, and place them into what has been called a "high-dimensional vector space," in essence internal hidden units of the algorithm, wherein analogies and similarities can be "recognized" and where the deeper, more complex analyses are made in ways that are not necessarily apparent to the user. This level of analysis *is not explicitly programmed by a person* who teaches the computer what to look for, nor has anyone programmed the computer as to the meaning of the data that the algorithm is "seeing" in its inputs; all of this "learning" is something the network does for itself. It is said to "learn" by means of creating and using its own internal criteria on which to make decisions, based on features of the training dataset that may or may not be apparent to the people who provide the training data. The system then uses its own internal criteria to make judgments on future data that are presented. In this way, an algorithm can identify objects on an image based on dimensional analyses that are not apparent to human eyes and minds. Sometimes their accuracy in doing so is astonishing, but their mistakes can also be somewhat bewildering, such as when a DL algorithm that was trained to classify general-content images can accurately name all the animals in photos of farms but interprets the photograph of a person as an "oboe."

For radiology, ML and DL applications are primarily focused on what are known as *classification problems*, wherein we want the algorithm to be able to answer specific questions based on imaging findings, such as "Is there a tumor present on this set of images?" Or, going a step further, "Is this particular tumor seen on this image likely to be benign or malignant?"

In such applications, the "hidden layers" of the neural-network architecture are what serve as the "feature detector," and since the machine teaches itself, it is free to "decide" what constitutes a "feature" that is important for making its classification. A recognized risk is that the algorithm might fit the training data poorly (which is termed "under-fitting") or that it can fit the training data too closely, including both its "signal" and "noise" elements ("over-fitting"). In either case, the algorithm will ultimately perform poorly on real-world examples.

The subtype of artificial neural network that is considered best suited to radiological image interpretation is known as a "convolutional" network, a special class of feed-forward networks that parse an image based on its smallest features, usually based on mathematical filters applied to small regions of the image, which are then analyzed through multiple convolutional layers of logic architecture, similar to the illustration in Figure 6.2. These can be used for all types of ML, including DL. The details of how this actually works in practice can get rather confusing, involving self-adjusting weighting of the connections between the many nodes, application of various subtly differing types of discriminant mathematical "activation functions" at each node, and the development of "feature vectors," which are mathematical constructs that guide machine decision-making through

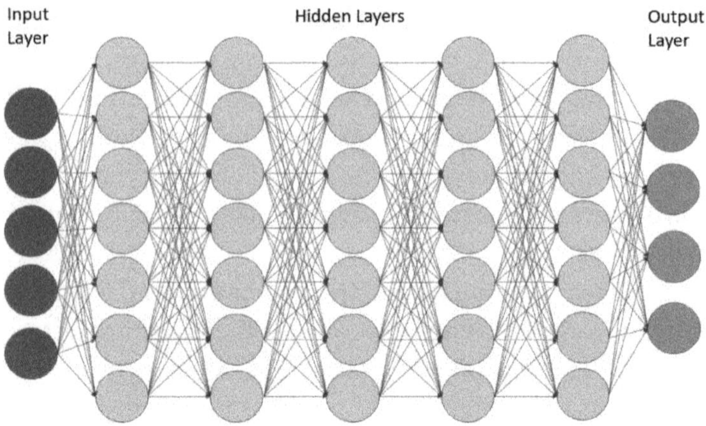

FIGURE 6.2. A complex, multilayered neural network with many hidden processing layers. This network is much more complex than the one illustrated in Figure 6.1 but is based on the same principles; as computers have become faster and more powerful, constructing artificial neural networks with higher levels of complexity is more feasible. These deeper networks are much more capable than their simpler predecessors. Image credit: Barry Amos, M.D., Penn State University.

highly iterative processes of progressive optimization. Inputs are assigned “weights” by means of mathematical “biases,” which are adjustable and which give some inputs greater influence over the final result than others. The “single processing unit” collects these weighted inputs and performs its mathematical function, resulting in an output that is “fed-forward” to multiple other similar processing units. In these systems, the number of actual computations needed to accomplish even the simplest image classification is extraordinarily large, so modern explosive increases in computer hardware speed and memory capacity have been essential for making these types of computer software applications feasible. Rather than wander into the weeds on the various detailed technical aspects of these software algorithms here, the interested reader is directed to the resources listed at the end of the chapter.[4]

An important point regarding DL, which is perhaps the most promising ML approach for carrying out complex analytical tasks, is that it is essentially a "black box": the details of how the algorithm accomplishes its learning and its decision-making tasks are *fundamentally unknowable*. In fact, the people who create the DL models themselves cannot even begin to explain how the algorithms do their work! DL models are said to be inherently, and fundamentally, opaque to the user. Remember, these are computers that, in essence, *program themselves* to perform the tasks they are set to, and the "reasoning" by which they accomplish this is deeply embedded (thus the term "deep learning") in the complex, self-adjusting behavior of large webs of "artificial neurons." It may be possible for a user to analyze a small, simple example of such a network and figure out what it must be doing, but once the network becomes large and multilayered, which is needed in order to have the power to make the sorts of discriminations that we want it to do in radiology (see Fig. 6.2), it also becomes fundamentally nonunderstandable. And it cannot explain itself, any more than human experts can truly explain how they accomplish their work.

This limitation has fueled new research aimed at producing "explainable AI," notably at the Defense Advanced Research Projects Agency (DARPA), a military research agency that also funded the research that led to the creation of Apple's Siri. As military interest in developing "intelligent" autonomous weapons, such as robotic tanks and weaponized drones, has developed, there are understandable concerns as to how these lethal devices might go about choosing their targets and carry out their functions, for example, so as to limit the kind of casualties the military likes to term "collateral damage" (e.g., noncombatants and other people who weren't intended to be killed). One wishes that such systems' decision-making can be made consistent with the rules of combat engagement, not to mention societal norms and ethics (e.g., that the systems will target only enemy combatants and refrain from killing surrendering soldiers, civilians, children, their pet goats, and so forth). Ultimately, it is quite difficult to trust a system whose workings are too complex to be explained, and this reality may ultimately limit the use of DL algorithms in areas that are considered to be too high risk, such as medical diagnosis, piloting airliners, and fighting wars. Presumably the practice of medicine in general, and diagnostic radiology in particular, falls into this general category.

To address this concern, AI experts have been working on finding ways for ML systems to provide their users some insight into the rationale for their outputs, such as displaying examples from their training dataset (e.g., keywords or parts of an image) that were the most significant in the computer model's decision-making process. But such insight is necessarily incomplete, and a great deal will always remain unexplained. Understandably, if defense researchers are trying to create autonomous weapons, such as "intelligent tanks" or self-directed lethal drones, the presence of hidden biases "built in" to the algorithms is a large and immensely important problem.

Biases that are inadvertently built in to (or that develop autonomously within) these algorithms are a significant problem for medical applications as well—and typically these biases are instilled inadvertently by the users' selection of the content of the training set, often in ways the users have not anticipated at all. For example, in one classic ML model developed for the US Army with the goal of using AI to detect camouflaged enemy tanks, the training set included 100 landscape photographs, 50 of which included military targets (enemy tanks) and 50 of which did not. This was a supervised learning set; the algorithm was told which images contained the target tanks. The goal was for the algorithm to accurately classify which were photos of landscapes with versus without military targets (enemy tanks) included. However, the engineers doing this work hadn't noticed that all of the "no target" training images were photos taken in bright sunshine and all of the "target" images were photos obtained on the same cloudy day. In the end, the algorithm very accurately learned to identify pictures that were obtained on cloudy days, and no other feature of the image had much impact on the resulting classification. In essence, the algorithm would shoot at anything it saw on a cloudy day and would never fire on an enemy tank if it saw one on a sunny day!

WHAT AI CAN DO, WHAT IT MIGHT DO IN THE FUTURE, AND WHAT IT CANNOT EVER DO

At this juncture, it is worth reviewing what AI can and cannot do, and comparing and contrasting that to people's hopes and aspirations for AI as it impacts the field of diagnostic radiology (Table 6.1). There have been a few notable successes of the DL approach in radiology that have been reported in the scientific literature. In 2017, David Larson, Curt Langlotz, and their group from Stanford University's Department of

Table 6.1

THINGS AI CAN AND CANNOT DO	
Things AI Can Do (Currently) or Might Do (Future)	*Things AI Cannot (Ever) Do*
Multiparametric analysis	Care about the patient
Image segmentation	Place decisions into ethical contexts
Region-of-interest delineation and delimiting	Contemplate its own biases
Detection of certain findings	Desire to improve itself
Incorporation of pretest probabilities	Recognize its own errors[5]
Perform statistical analyses	Overcome its own biases[6]
Image optimization	Empathize
Radiomic analyses of "hidden" features	

Radiology reported successfully training a DL algorithm to perform bone-age estimates from hand radiographs, which it was able to do to the level of 95% agreement with a panel of three expert human observers. Kheiron Medical Technologies, a UK-based company that collaborates closely with academic radiologists and imaging scientists in Europe, reported in 2018 that their DL algorithm achieves similar results for breast cancer screening using mammograms.

One surprising benefit of ML in radiology is that it may allow CT scans to be completed using much lower radiation doses than are currently needed; by the same token, ML may allow magnetic resonance imaging (MRI) scans to be done in a fraction of the time that they currently are. This is on the basis of the algorithms' potential ability to interpolate what is missing from noisy images and "sharpen" them by adding in the missing data, so that the resulting AI-derived images (partly based on patient data and partly estimated from probabilistic methods) are more easily interpretable by radiologists. The basic idea to this approach is that, with a lower CT radiation dose or with faster MRI scanning times, the data needed to fully form a high-quality image are simply missing. But the algorithm can "learn" from training sets what is most likely to have been lost in such a situation and then replace it in the final image product. This is currently an area of active research; since the computer would, in essence, be making an "educated guess" based on statistical probabilities, there is reasonable concern that this could introduce new sources of error in radiology. But if it can be made sufficiently accurate and reliable, this could be a boon to the field, from the standpoints of patient safety and patient access to imaging services.

Another potential application is for a DL system to "learn" what sorts of features on images matter to radiologists and then modify the images to provide attentional guidance to the human observer. This attentional guidance or "salience maps" can be presented in terms of colorized displays superimposed upon the grayscale medical image in order to subtly (or not-so-subtly) direct radiologists' attention to areas on an image where they are likely to find abnormalities. Such methods may have a large impact in reducing the most common type of radiologist errors.

As parallel diagnostic methods gain in depth and sophistication (e.g., the growing sciences of genomics, proteomics, metabolomics, and so forth as related to an expanded understanding of disease states), it is possible that even larger datasets correlating these factors to radiological appearances (known as "radiomics") will become available to train such algorithms. Radiomics is one of the most exciting potential applications of AI in radiology. The basic concept is that there may be discoverable imaging features of disease that, when present, can be used to infer specific phenotypic characteristics of an individual patient's disease process; this, in turn, might predict that particular patient's susceptibility to certain diseases or even his or her response to any particular treatment. Those imaging features that are reliably detectable by human observers today, such as the marginal characteristics of a detected lesion, or particular patterns of calcification in a soft-tissue mass, have been well described, and radiologists use them routinely. DL methods could extend this type of analysis considerably and may potentially gain the ability to identify myriad new radiomic features on images, including many that are actually imperceptible to humans, and thus may be able to make classifications of patients' disease states—including features that may never be readily accessible to human observers—based on "deep-learned" criteria from higher-dimensional analysis.

An example of this from the recent literature is a paper published in 2017 in the journal *Nature Medicine* by Drysdale et al., a group of scientists (including radiologists) at the Weill Cornell Medical Center. A DL algorithm that was trained on functional brain MRI images of patients suffering from depression was ultimately able to identify several subtypes of these depressed patients and to successfully predict which individuals might respond to a particular treatment (neurostimulation therapy).

It is also possible that these methods could drive basic science discoveries by revealing features of images that physicians do not currently realize have any particular clinical significance or predictive value. Finding unexpected or otherwise unobserved features in datasets is sometimes referred to as discovery science, in that it is serendipitous and unexpected, unlike hypothesis-driven science, in which experiments are designed to test preconceived ideas (although the two often overlap).

HOW IS MODERN AI BETTER THAN PRIOR FAILED ATTEMPTS AT CAD?

As we discussed in prior chapters, it is widely held that previous attempts at CAD in radiology have failed. Interest in CAD to aid radiologists in diagnosis has been ongoing at least since the first journal article on the topic appeared in 1994, testing a prototype CAD system for its ability to detect microcalcifications on mammograms. CAD systems have been in widespread clinical use, almost entirely in mammography, since the late 1990s (the US Food and Drug Administration approved CAD in mammography in 1998, and systems proliferated due to government payment subsidization). Yet most radiologists believe that commercially available CAD actually adds no real value to their basic task, with the CAD mostly flagging false positives or lesions that would not have been missed by the human observer. Careful research on the clinical effectiveness of CAD confirmed the users' impressions and showed that CAD had no impact on radiologists' accuracy, nor did it affect radiologists' decision making in mammography.

It is worth noting that "old CAD" failed even though it used sophisticated convolutional neural networks, the hallmark of modern DL. In the past, however, owing to limitations in computer processing power, the number of internal "hidden layers" of networked nodes was much more limited than it is currently. With increased computer power, as predicted by Moore's Law, the ability for true "deep learning" in such a system is greatly increased, allowing for more deep/hidden layers to exist and to "run" in a reasonable amount of time to be clinically useful, adding to the overall system's discriminating power.

Accordingly, the anticipated performance of these CAD "image classifier" systems should improve dramatically in the years ahead—that is, with a greatly reduced rate of false positives and false negatives, and with much more reliable true positives and true negatives. Another limitation of CAD was that it was trained in a "supervised leaning" model, using training image sets with known pathology, allowing trainer bias—in other words, they were not true DL in that the algorithms were not allowed to train themselves. It is thus likely that future CAD approaches using DL and modern computers will do much better, including allowing the evaluation of multiple prior studies over time and correlation with larger sets of clinical data that can be used to inform the classifier, none of which were possible previously. So it is not far-fetched to think that "new and improved" versions of CAD could soon become part of the radiologist's daily armamentarium.

In some cases, such a "super-CAD" might reasonably be expected to provide a substantial boost to radiologists' accuracy and efficiency. High-powered convolutional neural-network systems such as IBM's Watson could, in the future, essentially provide patients with a computerized "second reader" for every radiological study. "Old CAD" was somewhat oversold as being this electronic second reader and proved to be a huge disappointment when widely deployed in practice, but a future "super-CAD" might actually deliver on the original promise. Human second-reading has been shown to be enormously effective but is currently not feasible in clinical radiology on a useful scale, due to prohibitive costs and simply to limitations in the number of practicing radiologists vis-à-vis the large number of examinations that need to be interpreted. A new "super-CAD," in the form of a convolutional neural network that is highly sophisticated and that is trained effectively using DL methods, however, may provide a rapidly deployable, "clone-able," and cost-effective alternative electronic second reader. The benefits of second-reading have been discussed in prior chapters; in some studies, double-reading may possibly have led to as much as a 12-fold benefit in error prevention. Similarly, this CAD/human double-reader approach might well be expected to dramatically reduce radiologists' baseline error rate, especially for the perceptual errors that constitute the bulk of these and that have otherwise proven intractable to remediation. Recent research combining traditional CAD systems and DL methods has shown significant improvements in test performance (based on receiver-operator curve analysis), and it is likely that these systems will continue to improve at an accelerating rate.

Another drawback of the DL approach for computer-assisted detection and diagnosis is that, in order to succeed, such a system would require an *extremely* large number of medical images from which to "learn," and to be sure it is learning the right things, all of those images would have to be "correctly labeled," or else the algorithm will "learn" the wrong things, would acquire biases in its interpretation, and in the end would merely provide a highly sophisticated computerized error-propagation scheme.

This is probably why, for example, IBM acquired the Merge database of medical images for $1 billion in October 2015, creating the Watson Health Medical Imaging Collaborative. The Merge database held an extremely large number (tens of billions) of clinical medical images and associated clinical data, which had been acquired across 16 health systems, including academic medical centers and outpatient imaging centers. That acquisition effectively provided Watson with as many as 30 billion annotated radiological images with which to train its DL algorithm for medical imaging applications, as well as associated results of innumerable laboratory results, clinical notes, genomic tests, and related data stored in the electronic medical record data—data that serve as the "image labels." The stated goal is for Watson to be able to extract "invisible" insights from unstructured image data, to aid in detection and diagnosis and in radiomic classification of disease. No breakthroughs have yet been announced, and it is not yet even known if 30 billion radiological images are sufficient for training such a system to achieve the desired level of accuracy, but this ambitious project is ongoing.

AI VERSUS HUMAN ERROR IN RADIOLOGY

If you've read this book from Chapter 1 until this point, I think you will be inclined to agree that human beings can never be made error-free; it is simply not in our nature. Likewise, based on recent experience and an understanding of the complex nature of convolutional neural networks, the "black box" of DL, the sensitivity to biases and errors in the training samples, and also basic chaos theory, and so forth, it is also extremely likely that AI can never be made completely error-free, either. But it seems logical to think that the errors made by people would perhaps be altogether different ones than those errors likely to be made by machines. As a result, the idea of putting humans and machines together in a "symbiotic" functional unit may be the strategy that finally achieves the 70-year-old goal (i.e., since the time of Garland) of meaningfully reducing error in radiology below its stubbornly impenetrable 3% to 4% "floor," as well as improving the fraction (currently approximately 16%) of breast cancers on screening mammograms that are missed by radiologists, as well as other well-documented rates of error, including false-positive results on screening examinations (screening examinations being characterized by a very low prevalence of true positives) that are all likely due to limitations of the human observer, as we have discussed in prior chapters. It is altogether conceivable that an expert human observer, such as a trained and experienced radiologist, working in combination with a well-trained DL system (such as a future version of IBM's Watson), could yield results that are substantially better than either machines or humans could ever hope to accomplish on their own.

RADIOLOGY'S "BRAVE NEW WORLD"

In response to Professor Geoff Hinton's statement that machines will soon replace radiologists, an alternative future has been succinctly stated by thought leaders such as Drs. Curt Langlotz of Stanford University and Saurabh Jha of the

University of Pennsylvania: AI will almost certainly not replace all human radiologists, but those radiologists who can effectively use AI tools might eventually replace those who do not (or cannot).

Dr. Jha, a radiologist, and Eric J. Topol, a cardiologist who is the founder of the Scripps Translational Science Research Institute in La Jolla, California, published a brief paper in the *Journal of the American Medical Association* in December 2016 that has become somewhat of a focal point in the discussion of the future of AI in radiology. Entitled "Adapting to Artificial Intelligence: Radiologists and Pathologists as Information Specialists," Drs. Jha and Topol envision a near-future following the "fourth industrial revolution"—that is, after the rise of big data and AI revolutionize the workplace, augmenting or replacing human intelligence in myriad ways. In their view, the specialties of diagnostic radiology and pathology will be changed to a greater degree than other fields of medicine.

In their provocative two-page essay, Jha and Topol note that the amount of data derived from both imaging and laboratory testing is already quite large and continues to increase and will soon go beyond what can reasonably be extracted by (increasingly fatigued and overwhelmed) human radiologists and pathologists. Thus, computers and their software will increasingly be needed to take over the "data extraction" functions that are currently performed by these physician specialists. To avoid being entirely *replaced* by computers that are increasingly able to interpret images accurately and efficiently, Jha and Topol advise that radiologists and pathologists should "embrace AI" and allow themselves to be *displaced*—yielding to machines those tasks done better by machines and instead focusing their efforts on the *synthesis of information* that has been extracted by AI from scans and pathology images. The essence of their argument is that the radiologist of the future will no longer be primarily an interpreter of medical images but rather an "information specialist," and they recommend that the specialties of diagnostic radiology and pathology should perhaps be merged into a single new entity devoted to information synthesis tasks, guiding clinicians' decision making. The "information specialist" of the future, in their view, the descendent of present-day radiologists, will become the primary user of AI technologies and will use them to create greater benefits for patients than either AI alone or human beings alone could provide.

Is this the true future of the specialty of radiology and the future role of the diagnostic radiologist? As the rate of change in computer science is itself *accelerating*, it seems likely that the future state of diagnostic radiology (whatever it may be) will probably arrive earlier than expected. Technologies and approaches that seem inchoate currently may well become a part of routine practice in the next few years.

A paradigm shift of this magnitude is not without peril for those involved, including shifts in manpower needs with resultant economic consequences; for instance, won't fewer such "information specialists" be needed than the current number of practicing radiologists and pathologists? Some of us, no doubt, will experience a profound sense of loss if most of our past training and experience becomes essentially irrelevant to our work, and the tasks that previously filled our workdays and represented our personal contributions to patient care are changed into something new and (for some) completely different from anything we had previously imagined.

NOTES

1. On which I am honored to serve as a charter member.
2. Dr. Geoff Hinton is an emeritus professor of computer science at the University of Toronto and also an Engineering Fellow at Alphabet Inc.'s Google, where he leads the "Google Brain" AI team. At a 2016 conference on *Machine Learning and the Market for Intelligence* in Toronto, Canada, Prof. Hinton stated that it "seemed obvious" to him that DL would soon fully replace radiologists, saying: "I think if you work as a radiologist, you're like the coyote that's already over the edge of the cliff but hasn't yet looked down, so it doesn't realize that there's no ground underneath it. We should stop training radiologists now. It's just completely obvious that within 5 years deep learning is going to do better than radiologists." He went on to hedge a bit, saying "it might be 10 years; but we've got plenty of radiologists already."
3. An example: a recent experiment using this technique at my own institution discovered that, at least here in central Pennsylvania, patients treated in the hospital for congestive heart failure who recover and are discharged into ZIP codes where less than the average concentration of teenagers reside are (for unknown reasons) several times more likely to relapse and require readmission to the hospital. Unfortunately, the algorithm could only discover this strange association, it couldn't explain it.
4. Here is a link to an excellent online tutorial by Jason Brownlee in "Algorithms from Scratch" from the website "MachineLearningMastery.com," showing how to set up your own neural-network perceptron algorithm from scratch using the popular development package Python: https://machinelearningmastery.com/implement-perceptron-algorithm-scratch-python/ (accessed June 2018).

5. Perhaps debatable.
6. Not clear that humans are any better at this, however.

FURTHER READING

Dick P. *Do Androids Dream of Electric Sheep?* New York: Random House (Del Rey), 1968. ISBN 978-0-345-40447-3.

Drysdale T, Grosenick L, Downar J, Dunlop K, et al. Resting-state connectivity biomarkers define neurophysiological subtypes of depression. *Nature Med* 2017; 23(1):28–38. doi:10.1038/nm.4246

Erickson BJ, Korfiatis P, Akkus Z, Kline TL. Machine learning for medical imaging. *RadioGraphics* 2017; 37:505–515. doi:10.1148/rg.2017160130

Erickson BJ, Korfiatis P, Kline TL, Akkus Z, et al. Deep learning in radiology: does one size fit all? *J Am Coll Radiol* 2018; 15(3)B:521–526. https://doi.org/10.1016/j.jacr.2017.12.027

Jha S, Topol EJ. Adapting to artificial intelligence: radiologists and pathologists as information specialists. *JAMA* 2016; 316(22):2353–2354. doi:10.1001/jama.2016.17438

Jha S, Topol E. Information and artificial intelligence. *J Am Coll Radiol* 2018; 15(3)B:509–511. https://doi.org/10.1016/j.jacr.2017.12.025

Kahn CE. From images to actions: opportunities for artificial intelligence in radiology. *Radiology* 2017; 285(3):719–720.

Larson DB, Chen MC, Lungren MP, Halabi SS, Stence NV, Langlotz CP. Performance of deep-learning neural network model in assessing skeletal maturity on pediatric hand radiographs. *Radiology* 2017; 287(1):313–322. https://doi.org/10.1148/radiol.2017170236

McAleer S, Agosstinelli F, Shakov A, Baldi P. Solving the Rubik's cube without human knowledge. Available at: https://arxiv.org/abs/1805.07470v1; accessed June 2018.

Thrall JH, Li X, Li Q, et al. Artificial intelligence and machine learning in radiology: opportunities, challenges, pitfalls, and criteria for success. *J Am Coll Radiol* 2018; 15(3)B:504–508. https://doi.org/10.1016/j.jacr.2017.12.026

Topol E. *The Creative Destruction of Medicine: How the Digital Revolution Will Create Better Health Care.* New York: Basic Books, 2012. ISBN-13: 978-0465061839

Vonnegut K. *Player Piano.* New York: Random House (The Dial Press), 1952. ISBN-13:978-0-385033378-8.

Wilson DH. *Robopocalypse: A Novel.* New York: Random House (Doubleday), 2011. ISBN 978-0-385-53385-0.

EPILOGUE

Dealing with Our Mistakes

"I have M.D. after my name, and that means I don't MAKE mistakes!"

A 16-year-old girl with a history of lymphoma, now in remission following treatment, undergoes a routine surveillance annual positron emission tomography (PET)/computed tomography (CT) scan. The radiologist sees nothing amiss and comments that the patient's previously enlarged mediastinal lymph nodes remain small after a year's follow-up, with the largest measuring 9 mm. The voice-recognition transcription system she uses transcribes "9 cm" instead of "9 mm." Although she proofreads her report quickly, the radiologist does not notice this typographical error, and she signs and finalizes the report. The patient's father, anxious about the results, reads the radiologist's report on the online patient portal and becomes extremely alarmed and upset at the implications of a 9-cm lymph node. He immediately drives 90 minutes to the hospital to speak to his daughter's oncologist in person, phoning that doctor on his cell phone as he drives. The man is in extreme psychological distress as he arrives at the hospital and is greeted by his daughter's doctor, who has viewed the scan and saw no such 9-cm node. On contacting the radiologist, the typographical error is discovered and a correction is placed on the final report.

Unless a person is as self-deluded as my distinguished physician colleague who made the above statement without the slightest hint of irony, all of us—in every walk of life—will eventually have to deal with the realization that we have made a consequential error. Being wrong, making mistakes, and occasionally causing harm as a result of our errors are part of what it means to be human, whether or not you have graduated from a medical school.

If you are a radiologist, you will most often be confronted with the realization that you failed to see something of consequence on a radiograph or cross-sectional image, and it often seems obvious to you (and to everyone else) in retrospect. Most of us radiologists need only to reflect on our work over the past few days to think of an example. So it is worth discussing how we ought to deal with our mistakes when they are discovered.

Ideally, we deal with our mistakes in an open and forthright way, and at multiple levels: at the systems level, at the level of our practice organization, and importantly at a personal level as well. At the systems level we can perform root cause analyses when mistakes are discovered, and, with an eye on prevention, we can conduct failure modes and effects analyses prospectively, and apply the results methodically, with the aim of reducing variation and inconsistency in our practices so that our systems become more highly reliable and failures and systems errors more rare. We can conduct research to advance our knowledge of what various diseases look like on images, and find ways to teach radiologists at all levels more effectively, so that all patients benefit from the most current knowledge in the field regardless of where their studies are interpreted. At the practice-organizational level we can optimize our policies and procedures, better engineer our gadgets, and optimize our working conditions so that each person is able to perform optimally. At the individual level we can strive to be more self-aware, to avoid biases as much as possible, to reflect on our own practice and learn from our own mistakes as well as those of others, and also to strive to communicate more effectively with both the physicians we serve and the patients whose care we share with them.

But there is one other aspect of dealing with our own mistakes at the individual level that deserves greater emphasis: the need for all of us who have come to grips with the reality of our own mistakes to also accept the basic reality of our shared human fallibility, to place our human failings into the perspective of the broader human condition, and to find within ourselves a way to forgive ourselves for our own failings—and, by extension, to forgive our colleagues for theirs.

When we realize that we have made a mistake, we should recognize that our immediate reactions may be a little irrational. We may deny our error or repress its discovery, we may minimize or rationalize our error, and we may lie to ourselves and others about it. We may feel angry, as an element of our own self-concept may be threatened when our own errors are discovered. We may try to transfer the blame to others, even to the patient. Because our medical culture tends toward unrealistic expectations for perfection, we may feel shame, self-doubt, guilt, remorse, self-anger, or even a sense of outright self-loathing. But neither self-deception nor self-flagellation can serve any useful purpose. I would advocate a calmer, more dispassionate approach where impulsive responses are avoided—a "mental time-out," if you will—where we reflect on our mistake in its context, including the context of our emotional responses. It is only normal to feel badly when we find we have made a mistake, but it is how we deal with those emotions that matters most.

Some people have advocated performing a "cognitive autopsy" as soon as possible after the error is discovered. In this approach, set aside a quiet time when you are well rested and will not be interrupted. Sketch out a timeline of factors that led to the error, and try to mentally place yourself in the moment when the error occurred, with all of the uncertainty that was present then. Consciously try to avoid hindsight bias: think both prospectively and retrospectively about the error. At this stage, avoid discussion with others. Try to reframe the key clinical question as clearly and simply as possible. Only when you have completed this task should you discuss your thoughts with peers, and then re-review your analysis in the light of any objective external comments. Ask yourself what and how you were thinking at the time, and identify any cognitive errors that were made and their contribution to the outcome, if any. What systems errors came into play? Calmly and rationally consider what immediate actions may be needed—such as placing an addendum to correct errors in the written record, directly communicating with the referring physician, or making a formal error disclosure and apology to the patient. Financial compensation of some kind to harmed patients may also occasionally be appropriate.

Ultimately, however, if you want to continue in this line of work with optimal effectiveness and any degree of happiness and professional fulfilment, you must also try to find a path to self-compassion about your own human fallibility, let go of any grudge or anger you still hold against yourself, and find a way to forgive yourself for having made the error and return to your work with a clear mind, to do all you can to help as many sick and injured patients as effectively as is humanly possible in the years of your life that you have devoted to this work.

There is an emerging literature in the area of self-compassion, which is very distinct from self-pity or "self-esteem." Some particularly useful articles are included in the recommended reading list at the end of this chapter. Self-compassion is neither narcissistic nor self-delusional. It is not a "feel good" strategy as much as it is an acceptance of objective reality. It involves a focus on eliminating the grudge we bear against ourselves for being imperfect. It includes changing the "internal mental dialogue" inside our own minds so that it is less punishing. It involves acceptance of the human condition, the limits of which we are all subject to. It requires a large dose of humility, tempered with the realization that humility is not the same as humiliation. Self-compassion engenders a desire to improve oneself, as well as compassion for others, for we all share the human condition.

FURTHER READING

Neff, K. Self-Compassion: An Alternative Conceptualization of a Healthy Attitude Toward Oneself. *Self and Identity* 2:85-101, 2003.

Neff KD, Vonk R. Self-compassion versus global self-esteem: two different ways of relating to oneself. *J Personality* 2009; 77:23–50.
Neff KD, Rude SS, Kirkpatrick KL. An examination of self-compassion in relation to positive psychological functioning and personality traits. *Journal of Research in Personality* 41:908-916, 2007.
Schulz, K. *Being Wrong: Adventures in the Margin of Error.* Harper-Collins/Ecco © 2010 ISBN 978-0-06-117604-3
Longe O, Maratos FA, Gilbert P, Evans, G, et al. Having a word with yourself: Neural correlates of self-criticism and self-reassurance. *Neuroimage* 49:1849-1856, 2010.

INDEX

Page numbers followed by b, f, and t indicate boxes, figures, and tables, respectively.

www.ingramcontent.com/pod-product-compliance
Ingram Content Group UK Ltd.
Pitfield, Milton Keynes, MK11 3LW, UK
UKHW022023150926
475595UK00013B/256

9 780190 665395